MATLAB Applications in Engineering

Edited by Constantin Voloşencu

Published in London, United Kingdom

IntechOpen

Supporting open minds since 2005

MATLAB Applications in Engineering
http://dx.doi.org/10.5772/intechopen.91588
Edited by Constantin Voloşencu

Contributors
Bahadır Ergün, Cumhur Şahin, Septimiu Mischie, Nali Kumar Dinesh Kumar, Constantin Volosencu

First published in London, United Kingdom, 2022 by IntechOpen
IntechOpen is the global imprint of INTECHOPEN LIMITED, registered in England and Wales, registration number: 11086078, 5 Princes Gate Court, London, SW7 2QJ, United Kingdom
Printed in Croatia

British Library Cataloguing-in-Publication Data
A catalogue record for this book is available from the British Library

Additional hard and PDF copies can be obtained from orders@intechopen.com

MATLAB Applications in Engineering
Edited by Constantin Voloşencu
p. cm.
Print ISBN 978-1-83962-876-4
Online ISBN 978-1-83962-877-1
eBook (PDF) ISBN 978-1-83968-120-2

For EU product safety concerns:
IN TECH d.o.o., Prolaz Marije Krucifikse Kozulić 3, 51000 Rijeka, Croatia, info@intechopen.com or visit our website at intechopen.com.

Meet the editor

Prof. Dr. Constantin Voloşencu graduated as an engineer from Politehnica University of Timișoara, Romania, where he also obtained a doctorate degree. He is currently a full professor in the Department of Automation and Applied Informatics at the same university. Dr. Voloşencu is the author of ten books, seven book chapters, and more than 160 papers published in journals and conference proceedings. He has also edited ten books and has twenty-seven patents to his name. He is a manager of research grants, editor in chief and member of international journal editorial boards, a former plenary speaker, a member of scientific committees, and chair at international conferences. His research is in the fields of control systems, control of electric drives, fuzzy control systems, neural network applications, fault detection and diagnosis, sensor network applications, monitoring of distributed parameter systems, and power ultrasound applications. He has developed automation equipment for machine tools, spooling machines, high-power ultrasound processes, and more.

Contents

Preface

MATLAB is a computing program for engineers and scientists to analyze data, develop algorithms, and create models.

This book addresses specialists who are interested in the applications of MATLAB programs and technology. It covers some practical aspects of MATLAB programming, including new methodologies, techniques, and applications. New examples of MATLAB code and Simulink for technological developing platforms are promoted. These applications are from domains including electronic and communication engineering, geodetic and photogrammetric engineering, control systems, digital signal processing, and power electronics. Thus, the book discusses usages of MATLAB in radar antennas, geometric segmentation, Bluetooth applications, and control of electrical drives.

The published examples highlight the capabilities of MATLAB programming in the fields of mathematical modeling, algorithmic development, data acquisition, time simulation, and testing. Researchers in different domains developed new MATLAB applications and tools that enhance human understanding and improve specialist ability to design and implement high-performance solutions. Applications have presented that focus on the methodologies used, with implementation and testing issues.

The book is divided into four sections. The first section discusses mesosphere–stratosphere–troposphere (MST) radars, which are a type of wind profiler designed to measure wind speed and other atmospheric parameters up to altitudes of 100 km or more, characterized by high-power transmitters and large, very high-frequency antennas. The second section discusses geometric segmentation of geometric features into discrete geometric patterns that use real scanned geometries, noise-affected patterns, and that are not well sampled. The third section discusses applications of Bluetooth, which is a short-range, wireless technology standard used for exchanging data between fixed and mobile devices over short distances using ultra-high-frequency radio waves in industrial, scientific, and medical (ISM) purpose radio bands, excluding applications in telecommunications. The final section discusses modeling and simulation of electric drive control systems based on fuzzy PI speed regulators to improve control efficiency.

Chapter 1 is a brief introduction to the broad issues of MATLAB/Simulink applications, reviewing the areas to which the software is addressed and giving some basic examples of programming techniques. Chapter 2 gives an analysis to quantify the distortion in the radiation pattern due to aperture thinning at MST radar antenna. MATLAB is used to analyze the results of the radiation pattern, in both principal planed and for different azimuth angles with and without thinning, viewed in both polar and rectangular forms. Chapter 3 presents an application of automated building facade surveying produced from scanning data by means of coding in MATLAB. The chapter highlights that point cloud data need fundamental process flow as the fundamental steps of geometric segmentation by MATLAB programming. Chapter 4 presents low-energy Bluetooth applications in which measurement data are acquired by Bluetooth-compatible sensors and processed on a personal computer. The research application is using MATLAB elements such as methods to implement endless loops and real-time display of acquired data and using quaternions to handle the 3D orientation of a device. Chapter 5 presents a MATLAB program library for modeling and simulating speed fuzzy control systems for the main electric motors used as actuators in practice, including dc motors, induction motors, and permanent magnet synchronous motors.

I wish to thank the authors for their excellent contributions. I am also grateful to the staff at IntechOpen, especially Author Service Manager Dolores Kuzelj, for their assistance throughout the publication process.

Constantin Voloşencu
Department of Automation and Applied Informatics,
"Politehnica" University Timişoara,
Timişoara, Romania

Section 1

Introduction

Chapter 1

Introductory Chapter: Matlab and Simulink Applications

Constantin Volosencu

1. Generalities and publications

In the scientific and technical field there are a multitude of numerical calculation programs. Some examples of these programs can be given as follows. Analytica, created by Lumina Decision Systems, is a numerical modeling environment with a visual programming language based on influence diagrams. LabView, created by National Instruments, is a graphical and textual through formula nodes software, for process monitoring and control. Mathcad, created by Parametric Technology Corporation, is a computer software for the verification, validation, documentation and re-use of mathematical calculations. Matlab is a proprietary multi-paradigm programming language and numeric computing environment developed by MathWorks. It allows numerical computation and simulation with extended 2D/3D visualization with vector manipulation. Matlab allows matrix manipulations, plotting of functions and data, implementation of algorithms, creation of user interfaces, and interfacing with programs written in other languages. Simulink is a Matlab-based graphical programming environment for modeling, simulating and analyzing dynamical systems from different domains.

According to MathWorks presentation, the Matlab language fundamentals consists in syntax, operators, data types, array indexing and manipulation. Some of mathematics domain supported are linear algebra, differentiation and integrals, Fourier transform and other. The users may presents graphic results in two and three dimensional plots, images, animation and visualization. Data can be imported and exported, analyzed, preprocessed and visually explored. The language has many functions and assures programming scripts with program files, control flow, editing and debugging. The users may develop applications using App Designer, Guide or a programmatic workflow. Advanced software development is supported with object-oriented programming, code performance, unit testing, external interfaces to Java and Web services, C/C++, .NET and other languages. The language is a desktop environment with preferences and settings and platform differences. It assures support for third-party hardware, such as webcam, Arduino, and Raspberry Pi hardware. Simulink is a block diagram environment for multidomain simulation and model-based design. It supports system-level design, simulation, automatic code generation, and continuous test and verification of embedded systems. Simulink provides a graphical editor, customizable block libraries, and solvers for modeling and simulating dynamic systems. It is integrated with Matlab, enabling users to incorporate Matlab algorithms into models and export simulation results to Matlab for further analysis. It allows modeling of time-varying systems and of large-scale architectures, running systems, reviewing results, validate system behavior, optimizing performance for specific goals. The users may extend the existing Simulink modeling functionality using Matlab, C/C++, and Fortran code.

It assures hardware support for third-party hardware, such as Arduino, Raspberry Pi, and Beagle Board.

The software has many applications in practice, which can be mentioned: signal processing, image processing and computer vision, control systems, test and measurement, radio-frequency and mixed signals, wireless communication, robotics and autonomous systems, automotive, aerospace, FPGA, ASIC, and SoC Development, computational finance, computational biology and the number of application is increasing.

Over the years, numerous books have been published that present applications of the Matlab and Simulink programs. Some examples from the last years can be highlighted, as follows. Some books dedicated to students and engineers, which presents fundamentals of Matlab may be mentioned. This books make introduction in basic programming, fundaments consisting in data, statement structures, control structures, functions, algebraic computation, variables, complex numbers, vectors and matrices, data processing, examples of solving problems, examples in chemistry and physics, but also some advanced techniques for object-oriented programming, graphical user interface design and web applications [1–7]. More advanced issues as model predictive control or deep learning application are presented in [8] and respectively [9].

Extensive collections of works in the field of Matlab and Simulink applications, from the last years, can be cited as follows [10–20]. These collections, which can be used for educational, scientific and engineering purposes, include applications of: programming, developing graphical user interfaces, power system analysis, control systems design, system modeling and simulation, parallel processing, optimizations, signal and image processing, computer graphic visualization, electric machines, power electronics, genetic programming, digital watermarking, artificial networks, algebraic computation, data acquisition, image processing, seismology, meteorology, natural environment, interconnected power grids, antennas, underwater vehicles, models and data identification in biology, fuzzy logic, and discrete event systems.

Papers using the Matlab and Simulink programs have appeared and continue to appear in the literature. Here are some examples from the last years. A Matlab processing toolbox for analytical spectral devises field spectroscopy data, for generation of consistent and comparable ground spectra that have been corrected for viewing and illumination geometries as well as other factors such as the individual characteristics of the reference panel used during acquisition [21]. A software development platform is used in [22] for speedy evaluation and implementation of image processing options on the automatic guided vehicles. A program code written in Matlab, designed to be used inside of a Simulink model in [23], allows a fuel cell model to be used in a wide variety of 1D simulation platforms by exporting the code as C/C++.

2. Examples

2.1 A hyperbolic partial differential equation

The following example is realized using the *PDE modeler* toolbox. With this application the users can analyze elliptic, parabolic and hyperbolic Eqs. A hyperbolic equation case study, for wave propagation in square domain in plane, is presented in this example [24, 25]. The equation used in analysis is:

$$\frac{\partial u^2}{\partial^2 t} = c_1 \nabla (c_2 \nabla u) + c_3 u + c_4 \tag{1}$$

where the parameter have the following values: $c_1 = 1$, $c_2 = 1$, $c_3 = 0$, $c_4 = 10$.

The space on which is made the analysis is a square with unitary dimension $l = 1$. Boundary conditions were imposed as follows: on the left, right and front Dirichlet conditions: $h = 1$, $r = 0$. On the square's base Neumann conditions: $q = 0, g = 0$.

The discrete optimized number and position of meshes are presented in **Figure 1**. The contour solution is presented in **Figure 2**. For these meshes the approximated solution is presented in 3D in **Figure 3**.

2.2 Modeling and simulation of a control system for a second order process

The second example presents a simple case of modeling and simulation of a basic control system for a second order process.

The process has the transfer function:

$$H_p(s) = \frac{y(s)}{u(s)} = \frac{K_L}{(T_1 s + 1)(T_s + 1)} \tag{2}$$

where u is process input and y is process output.

The following values are chosen for the process parameters: $T_1 = 0{,}4$ s, $T_s = 0{,}04$ s and $K_L = 2$.

The process has a disturbance v at its input.

A linear PI controller is chosen, with the transfer function:

$$H_R(s) = \frac{u_c(s)}{e(s)} = K_R\left(1 + \frac{1}{T_i s}\right) \tag{3}$$

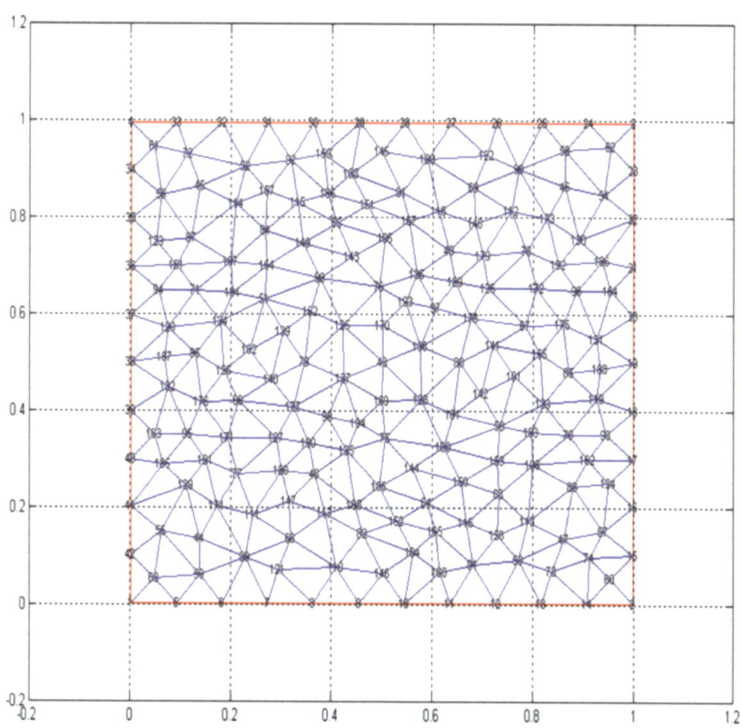

Figure 1.
The optimized meshes.

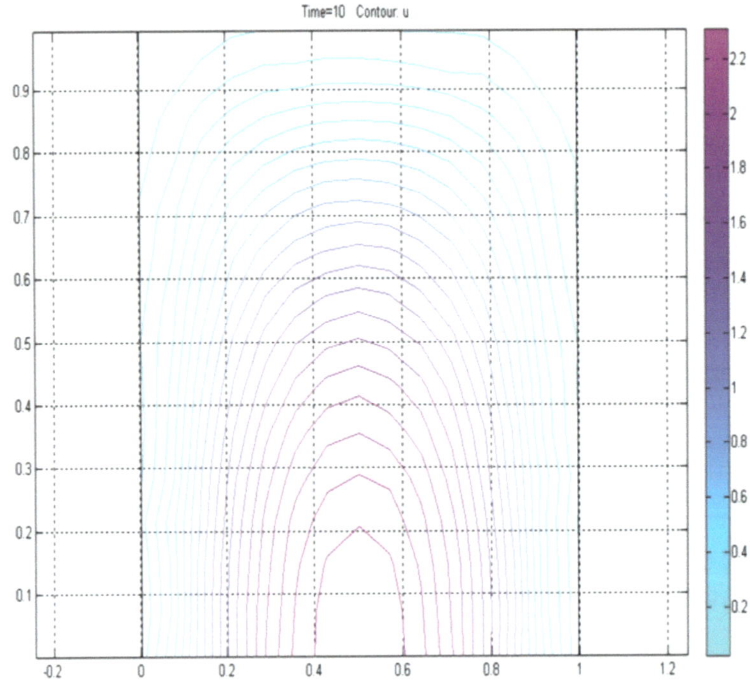

Figure 2.
Contour plotted solution.

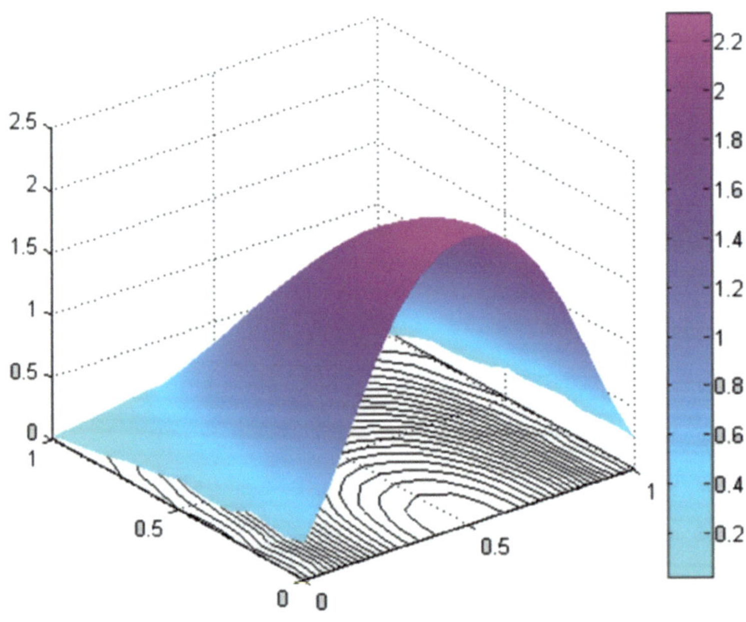

Figure 3.
Plotted solution in 3D.

where e is the error, as difference between the reference w and the feedback r and u_c is the command. The controller is tuned in accordance with the Kessler version of the module criterion:

$$T_i = T_1 = 0,4$$
$$K_R = \frac{T_1}{2K_L T_s} = 2,5 \tag{4}$$

In **Figure 4** shows how to arrange the work windows for this application on the screen. First, Matlab work space is open, then Simulink. A Simulink block diagram for the control system according the above theory is developed, with *transfer function* and *integrator* blocks from *Continuous* block library, *Add* and *Gain* blocks from *Math Operators*, *Step* and *Clock* blocks from *Sources*, *To Workspace* and *Scope* blocks from *Sinks*. The parameters are entered literally in the Simulink scheme, and their values are given in Matlab. The parameter values are saved in a data file, which is called with instruction *load* each time before the scheme is run. The values of the vectors w, uc, v, u, y, calculated at the tome values from vector t, are passed into Matlab workspace with the blocks *To Workspace* The time variations of the variables

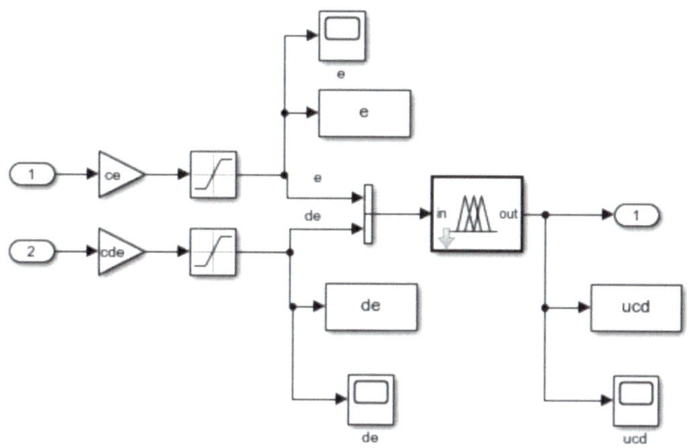

Figure 4.
Simulink block diagram work screen.

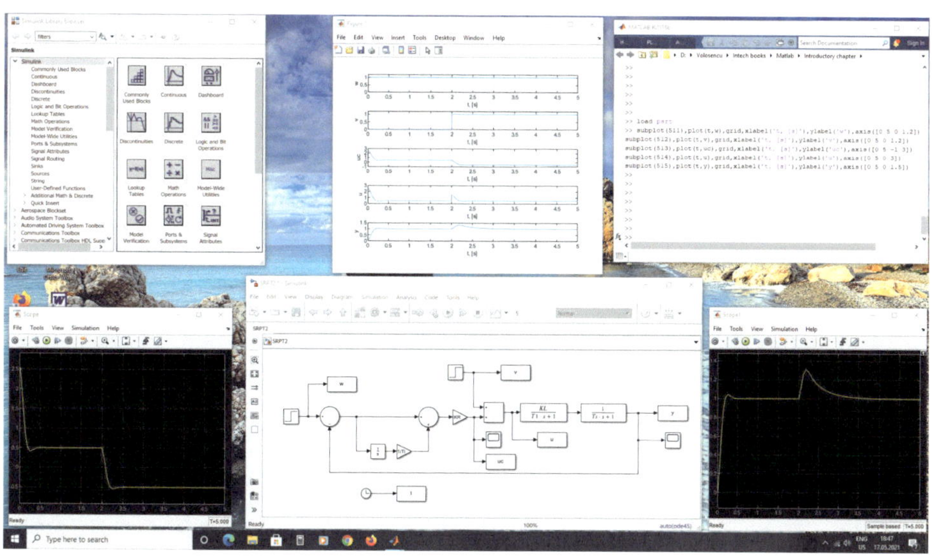

Figure 5.
Simulink diagram for fuzzy block.

u_c and y are presented on the two scopes. The time variations of the variables w, v, u_c, u and y are presented using the instructions *suplot*, *plot*, *grid*, *xlabel*, *ylabel*, *axis*.

Analyzing the output graph y it can be seen that the overshoot $\sigma_1\% = 4{,}3\%$, the settling time $t_r = 8{,}4.T_s = 0{,}336$ s, in accordance with Kessler's tuning criterion.

2.3 Modeling and simulation of a fuzzy control system for the second order process

The third example presents a simple case of modeling and simulation of a basic fuzzy control system [26–29] for the same second order process like in the second example.

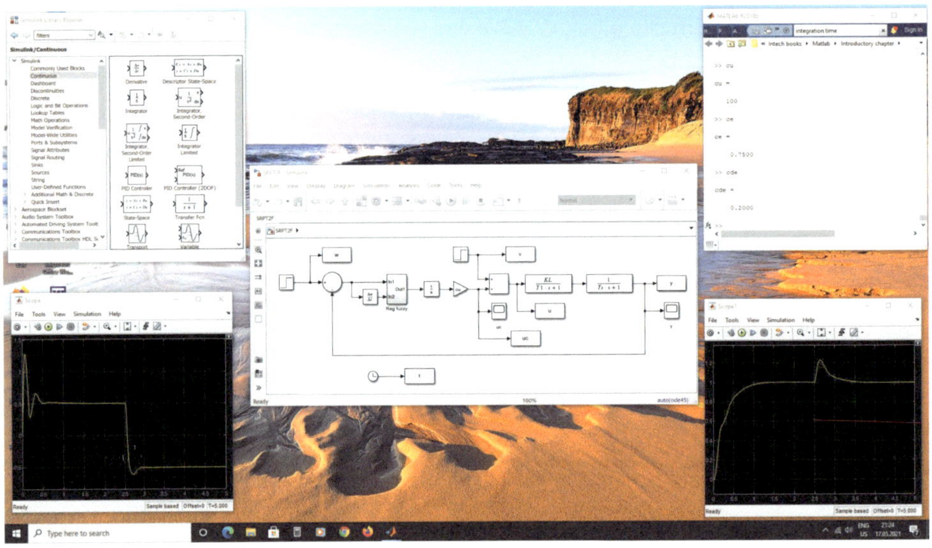

Figure 6.
The screen for fuzzy control system.

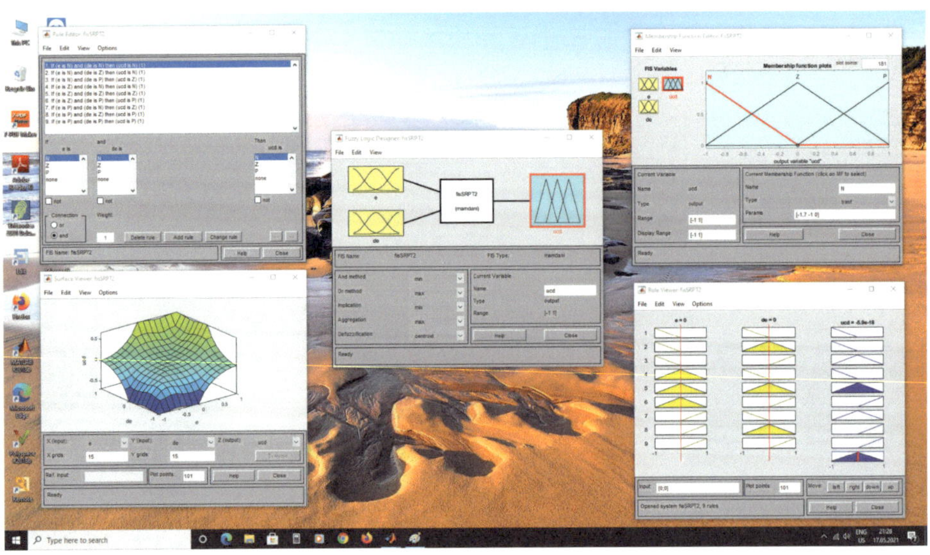

Figure 7.
The screen for fuzzy system design.

The process has the same transfer function. A fuzzy PI controller is used.

In **Figure 5** shows Simulink block diagram for fuzzy controller. With the *fuzzyLogicDesigner* window the user may set membership functions, rule base and inference.

In **Figure 6** shows how to arrange the work windows for this application on the screen.

In **Figure** 7 shows the work window for *fuzzyLogicDesigner*. The Simulink block diagram uses a fuzzy controller with derivation at the input and integration at the output. The working with Matlab and Simulink is the same like in the second example.

Conflict of interest

The author has no conflict of interest.

Author details

Constantin Volosencu
"Politehnica" University, Timisoara, Romania

*Address all correspondence to: constantin.volosencu@aut.upt.ro

References

[1] Chapman S.J. Matlab Programming for Engineers, 6th edition, Cengage Learning, 2020.

[2] Xue D. Matlab Programming. Mathematical Problem Solutions, De Gruyter, 2020, DOI:10.1515/9783110666953.

[3] Hahn B., Valentine D. Essential Matlab for Engineers and Scientists, Academic Press, 2019.

[4] Lee H.H. Programming and Engineering Computing with Matlab, SDC Publications, 2019.

[5] Moore H., Matlab for Engineers, 5th Edition, Pearson, 2018.

[6] Attaway S. Matlab 5th Edition A Practical Introduction to Programming and Problem Solving Butterworth-Heinemann, 2018.

[7] Kattan P.I. Matlab For Beginners: A Gentle Approach, Createspace, 2008.

[8] Dittmar R. Model Predictive Control mit Matlab und Simulink, IntechOpen, London, UK, 2019, DOI: 10.5772/intechopen.86001.

[9] Paluszek M., Thomas S. Practical Matlab Deep Learning: A Project-Based Approach, Apress, 2020.

[10] Leite E.P., editor, Matlab Modelling, Programming and Simulations IntechOpen London UK 2010 DOI: 10.5772/242.

[11] Assi A., editor, Engineering Education and Research Using Matlab IntechOpen London UK 2011 DOI: 10.5772/1532.

[12] Ionescu C., editor, Ed. Matlab A Ubiquitous Tool for the Practical Engineer IntechOpen London UK 2011 DOI: 10.5772/82.

[13] Leite E.P., editor, Scientific and Engineering Applications Using Matlab IntechOpen London UK 2011 DOI: 10.5772/1531.

[14] Chakravarty S., editor,. Technology and Engineering Applications of Simulink IntechOpen London UK 2012 DOI: 10.5772/2414.

[15] KatsikisV., editor, Matlab A Fundamental Tool for Scientific Computing and Engineering Applications - Volume 1, IntechOpen, London UK, 2012, DOI: 10.5772/2557.

[16] Katsikis V., editor,. Matlab A Fundamental Tool for Scientific Computing and Engineering Applications - Volume 2 IntechOpen London UK 2012 DOI: 10.5772/3338.

[17] Katsikis V., editor, Matlab. A Fundamental Tool for Scientific Computing and Engineering Applications - Volume 3, IntechOpen, 2012, London UK, DOI: 10.5772/3339.

[18] Bennett K., editor,. Matlab. Applications for the Practical Engineer, IntechOpen London UK, 2014, DOI: 10.5772/57070.

[19] Valdman J., editor Applications from Engineering with Matlab Concepts IntechOpen London UK 2016 DOI: 10.5772/61386.

[20] Saghafinia A., editor, Matlab. Professional Applications in Power System, IntechOpen London UK 2018, DOI: 10.5772/intechopen.68720.

[21] Elmer, K.; Soffer, R.J.; Arroyo-Mora, J.P.; Kalacska, M. ASDToolkit: A Novel Matlab Processing Toolbox for ASD Field Spectroscopy Data. *Data* 2020, *5*, 96. DOI: 10.3390/data5040096.

[22] Kotze, B.; Jordaan, G. Investigation of Matlab as Platform in Navigation and

Control of an Automatic Guided Vehicle Utilising an Omnivision Sensor. Sensors 2014, *14,* 15669-15686. DOI: 10.3390/s140915669.

[23] Lazar, A.L.; Konradt, S.C.; Rottengruber, H. Open-Source Dynamic Matlab/Simulink 1D Proton Exchange Membrane Fuel Cell Model. Energies 2019, *12,* 3478. DOI: 10.3390/en12183478.

[24] Voloşencu, C., Identification of Distributed Parameter Systems, Based on Sensor Networks and Artificial Intelligence, WSEAS Transactions on Systems, 2008, Issue 6, Vol. 7, p. 785-801.

[25] Voloşencu, C. - Identification in Sensor Networks, In: Proceedings of the 9th WSEAS International Conference on Automation and Information (ICAI'08), June 24-26, 2008; Bucuresti, WSEAS Press, p. 175-183.

[26] Voloşencu, C., editor, Fuzzy Logic, IntechOpen Ltd., London, UK, 2020, DOI: 10.5772/Intechopen.77460.

[27] Volosencu, C., Introductory Chapter: Basic Properties of Fuzzy Relations, *Fuzzy Logic*, IntechOpen, London, UK 2020, *DOI:* 10.5772/Intechopen.77460.

[28] Volosencu, C., Tuning Fuzzy PID Controllers, *Theory, Tuning and Application to Frontier Areas*, edited by Rames *C. Panda*, InTech, 2012. DOI: 10.5772/32750.

[29] Voloşencu, C., Stabilization of Fuzzy Control Systems, WSEAS Transactions on Systems and Control, 2008, Issue 10, Vol. 3, p. 879-896.

Section 2

MST Radar

Chapter 2

Radiation Power Pattern Distortion Analysis Using MATLAB for MST Radar System

Nali Dinesh Kumar

Abstract

Most often, in MST radar system, a few number of transmitters are non-operational due to various factors, making the liner sub-arrays corresponding to these transmitters in effective. This results in the thinning of the aperture and deviation of the excitation from the specified Taylor distribution. The array pattern will be distorted due to this deviation, when compared to the reference pattern. This chapter gives a complete analysis to quantify the distortion in the radiation pattern due to Aperture thinning. MATLAB was extensively used to analyze the results. The results of the radiation pattern in both principal planed and for different azimuth angles with and without thinning/tilt are presented. Radiation pattern is viewed in both polar and rectangular (2-D and 3-D) forms. Conclusions on the results obtained are presented.

Keywords: Array Factor, Distortion, Aperture thinning, MATLAB, Phased antenna, Polar Form, Rectangular form, Side Lobe Levels, Taylor distribution

1. Introduction

The ever-increasing demand for the software development of Aperture thinned Radiation Pattern has motivated to model the present work. The Indian MST radar is a phased antenna array, highly sensitive operating at 53 MH_Z with a peak power aperture product of 2.5×10^{10} W-m^2. One for each polarization, it consists of two collocated orthogonal sets of 1024 3-element Yagi-Uda antennas. They are arranged over an area of 130 m $\times$ 130 m in a 32 $\times$ 32 matrix (**Figure 1**). The complete array setup is illuminated using 32 distributed transmitters of varying power. In turn distributed transmitters each will feed a linear sub-array of 32 antennas with a 32 parallel runs of center-fed-series-feed structures [1].

Yagi-uda antenna is chosen for MST radar antenna array. Choice of an element for MST radar, advantages in favor Yagi element, requirements of side lobe level (SLL), antenna element design and modified Taylor distribution are explained [1–4]. Amplitude distribution, illumination efficiency and feeder efficiency are derived. Finally MST radar specifications are tabulated.

While planning for the antenna element designing, the sharing antenna elements architecture of fixed overlap sub array to avoid grating lobe in the antenna pattern technique is also considered [5].

An experiment made to generate low side lobe patterns optimizing ring radii and individual element excitations from concentric circular arrays [6] does not worked

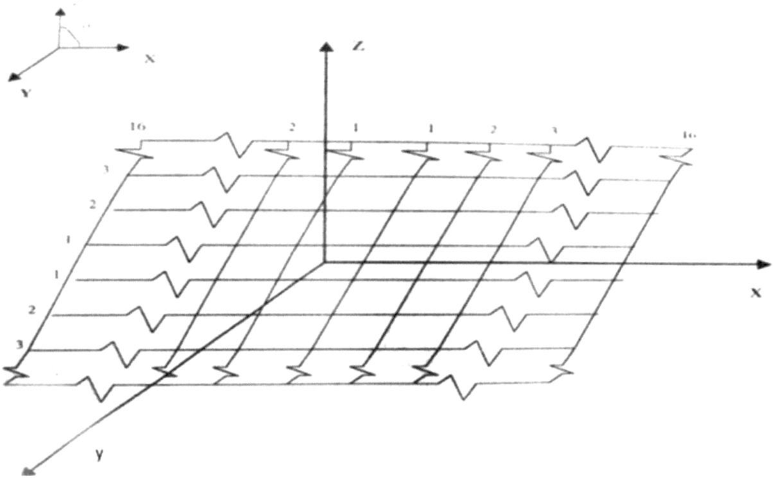

Figure 1.
Geometry for MST radar system.

for MST radar array. The approach of array excitation weight vectors as imaginary number chromosomes are often used as a general tool for pattern synthesis of absolute arrays that uses decimal linear crossover [7]. WWII elevated the phased-array antennas and has become a perfect tool for RF systems [8]. Then the main focus on the G-band multifunction measuring instrument systems for the Land is developed [9]. A coupled structural-electromagnetic model of phased array antenna PAA is developed to explain the performances of antenna, and the result of random errors and mechanical distortion [10]. Random error is generated throughout the producing and assembly method, and mechanical distortion is caused by external masses like high thermal distinction, vibration and impact loads [11]. Arrays produce aperture errors [12] as their determination is sometimes neglected, being in several cases very troublesome, such errors area unit mutual effects between parts of AN array, scattering from and obstruction because of the feed of a parabolic reflector, and optical phenomenon at a lens antenna step etc. [13].

2. Geometry of MST radar

Indian MST Radar antenna array uses a two-dimensional filled antenna array for both transmission and reception. An inter-element spacing of 0.7λ (λ, being the radar wavelength) is used in both the principle directions, which allows a grating lobe free beam scanning up to an angle of about 24^0 from broadside direction [1].

2.1 Choice of the element

To obtain the gain of 36 dB as given in the MST radar specifications a filled aperture of roughly 21λ to 25λ is required. To fill this aperture the number of elements required are given as

$$N = \frac{\text{Total Aperture Area } A_p}{\text{Effective Area of Single Element } A_e} \quad (1)$$

Where $A_e = \lambda^2 G_e / 4\pi$.

S.No	Types of Elements	Effective Gain G_e	Effective Area A_e	No. of elements N	Approximate grid spacing
1	Isotropic	1	λ^2 / 12	63	0.29
2	Dipole	1.5	λ^2 / 8	42	.35
3	Dipole over ground Plane	3	λ^2 / 4	21	0.5
4	Yagi 3 element	5	λ^2 / 2.5	12	0.64
5	Yagi 4 element	8	λ^2 / 1.5	79	0.83

Table 1.
Different types of elements, and their number required to fill the aperture.

Table 1 gives the total number of elements required to fill up the aperture for different types of antenna elements. Comparative study of the various antennas as the potential elements in MST radar configuration was made the possibility of using following types. They are Crossed Dipole over a ground plane, Coaxial Collinear, Three –element Yagi, and Four-element Yagi.

Out of these elements, the crossed dipole over a ground plane has the gain of the order of 5 dB and hence total number of dipoles required to fill the same aperture is quite large compared to the Yagi types and would require a more complicated and expensive feed network. The Coaxial Collinear *(CoCo)* antenna, which again is another form of dipole over a ground plane, is apparently simple to fabricate. These can be directly constructed at the site using RG 8/U or equivalent RF cables, but maintenance and water proofing of such an array would be tough.

A ***Yagi-Uda*** array consists of many parallel dipoles with different lengths and spacing, out of which only one is actively fed and others are shorted at their feed points. Since only one of the dipoles is driven and all other elements are parasitic, the later functions respectively as a reflector or as a director. In general, the longest shorted element with length of the order of λ/2 is the reflector and the shorted element is the director. This can be viewed as the array of dipoles in which all but the driven elements (Exciters) are short-circuited. For three-element Yagi case Voltage is

$$V_m = \sum_{n=1}^{3} I_n Z_{mn} \tag{2}$$

Where, I_n is the current on the nth element
m is the element number
Putting $V_1 = V_2 = 0$ and simultaneously solving these equations gives

$$\frac{I_1}{I_2} = \frac{Z_{13}Z_{32} - Z_{12}Z_{33}}{E} \tag{3}$$

$$\frac{I_3}{I_2} = \frac{Z_{13}Z_{12} - Z_{23}Z_{11}}{E} \tag{4}$$

Where, $\mathbf{E = Z_{11}\,Z_{33} - Z_{31}\,Z_{13}}$.
Using these current rations, the input impedance, gain and radiation pattern can be calculated.

2.2 Antenna element design consideration

The single element gain and radiation pattern change considerably in the array environment. The physical area that each element couples limits the element gain in an infinite array [2, 3, 14] and is given by

$$g_r = \left(4\pi\, d_x\, d_y/\lambda^2\right) \cos\theta \quad (5)$$

for, $d_x = d_y = 0.7\lambda$

$$g_r = 4\,\pi\,(0.49) = 6.157 = 7.89 \text{ dB} \quad (6)$$

A practical element with a gain higher that this value, would lead to overlap of effective areas, without any useful addition to the array gain. The three-element Yagi appears to be a practical choice as the element of the MST array. It has a higher front-to-back ratio, which is useful in minimizing the ground effects and it can be designed to have a gain between 6.5 dB to 8 dB.

Considering the isolated Yagi element gain as 7.2 dB, the total array gain at a taper frequency of 80% works out to be 36.3 dB. This would leave a margin of 0.3 dB towards gain loss due to amplitude and phase errors across the aperture, thus allowing us to realize a gain of 36 dB for the zenith beam. The diameter of the element was chosen to be 0.75 inch, which is a standard commercially available tube.

The following values were found to offer satisfactory performance

L1: Length of the reflector	=2.799 m.
L2: Length of the exciter	=2.677 m.
L3: Length of the director	=2.369 m.
D1: Distance between reflector and exciter	=1.245 m.
D2: Distance between director and exciter	=0.895 m.

The expected performance of thee-element Yagi with the above parameters is tabulated below

Gain	7.236
Input Impedance	52.75 - j14.3
Front-to-back ratio	15 dB

2.3 Feeder network configuration

The feeder network of MST radar antenna array consists of two orthogonal sets; one for each polarization. The feeder network consists of thirty-two parallel runs of center-fed-series-feed (CFSF) structure. Thirty-two transmitters of varying power illuminate the array; each is feeding a linear sub-array of thirty-two antenna elements.

The feeder networks of all the sub-arrays are identical as far as the power distribution is concerned. The CFSF network, (shown in **Figure 2**) consisting of power divider at the center and a series of directional couplers on each side of its, connects the linear sub-array to the T/R switch, which delivers the transmitter output power to the array and the power received by the array to the corresponding low noise amplifier. Components of the feeder network are RG 1–5/8″, RS 7/8″ and air dielectric coaxial lines, Wilkinson type in-phase power divider, Distributed version of coupled line type directional couplers and Lumped version of hybrid type directional couplers [1].

Description of each of the above components is given below.

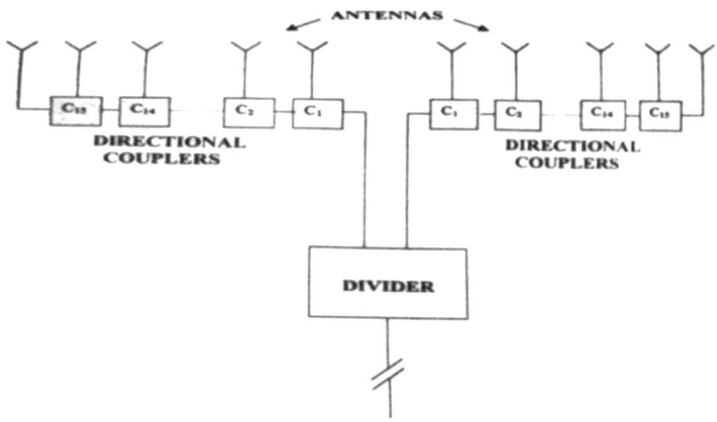

Figure 2.
Center -fed-series-feed (CFSF) network.

2.4 Rigid cable

RG 1–5/8″ cable is used to carry the output power of high power transmitters *(70 kW – 120 kW range)* to the CFSF network. RG 7/8″ is used to carry the output power of low power transmitters *(15 kW – 53 kW range)* to the sub-array input. This cables use foam dielectric, which is having a velocity factor of 0.89 at 53 MHz, the operating frequency, these cables offer an attenuation of about 0.5 dB per 100 m.

2.5 Power divider/combiner

This device acts as a divider in the transmit mode and as a combiner in the receive mode. The circuit diagram of *Wilkinson type divider/combiner* is shown in **Figure 3**. All the three ports are terminated with characteristic impedance, Zo (50 Ω). Ports 2 and 3 are isolated. During the transmit mode the transmitter output power is fed to the port-1, which will be divided in phase equally between the output ports 2 and 3. In the receive mode the power received by the two halves of the linear sub-array will be delivered through series feed network to the ports 2 and 3 respectively which will be combined in phase art port-1.

The relationship between the voltages in the transmit mode at the output and input ports is given by

$$V_2 = V_3 = -j\,(V_1/V_2) \tag{7}$$

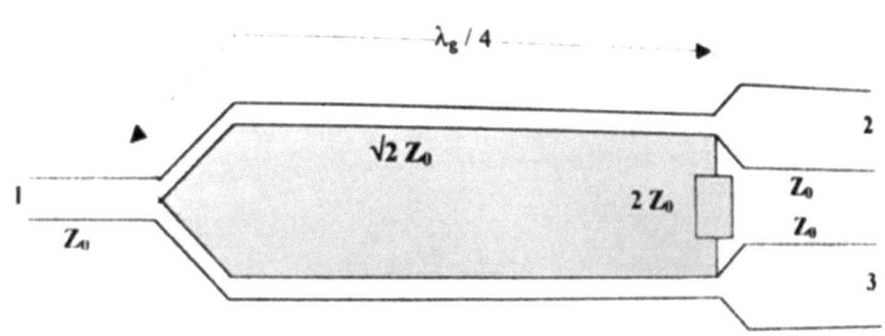

Figure 3.
Wilkinson type divider/combiner.

The relationship between the voltages in the receive mode at the output and input ports is given by

$$V_1 = -j\,(V_2/V_3)/\sqrt{2} \tag{8}$$

Where, V_i is the voltage at the port-I. Since the two halves are symmetry $V_2 = V_3 = V$. Therefore.

$$V_1 = -j\left(\sqrt{2}\right)V \tag{9}$$

2.6 Distributed versions of coupled line directional coupler (DC)

The coupled rod co-axial directional coupler is shown in **Figure 4**. In the **Figure 4**, section, 1–2 is the main line and 3–4 is the auxiliary line, which is coupled to the main line. As the electric current passes through Section 1–2 from port-1, it produces a magnetic field around it. This magnetic field couples with conductor 3–4 and induces current in it. Therefore, by varying the separation between the two conductors, we can control the coupling factor.

In the transmit mode port-1 is the input port to which the power will be fed. Port-2 is the direct output power and portr-3 is the coupled port through which antenna will be energized. Port-4 is isolated with respect to port-1. The relationship between various voltages is given by

$$V_3 = k\,V_1 \qquad V_2 = -j\left(\sqrt{(1-K^2)}\right) \times V_1 \qquad V_4 = 0 \times V_1 = 0 \tag{10}$$

Where Vi is the Voltage at port-i. This indicates that V_1 and V_3 are in phase and V_2 is lagging V_3 by 90^0.

All the thirty-two antennas within a sub-array should get the excitation signals with same phase so as to produce a main beam in the broad side direction resulting in high gain. In order to achieve this, the lengths of the feeding cables (running from the coupled port to the antenna balun) are adjusted accordingly. This process is called *Phase equalization*.

In the receive mode the antenna delivers power to the coupled port (port-3) and port-2 will be fed by the power coming from the adjacent coupler. In this mode, as a consequence of the phase equalization, V_2 always leads V_3 by 90^0. The relationship between the various voltages is given by

$$V_1 = \left(\sqrt{(1-K^2)}\right)V_2 + k\,V_3 \tag{10a}$$

$$V_4 = \left(\sqrt{(1-k^2)}\right)\,V_3 - k\,V_2 \tag{11}$$

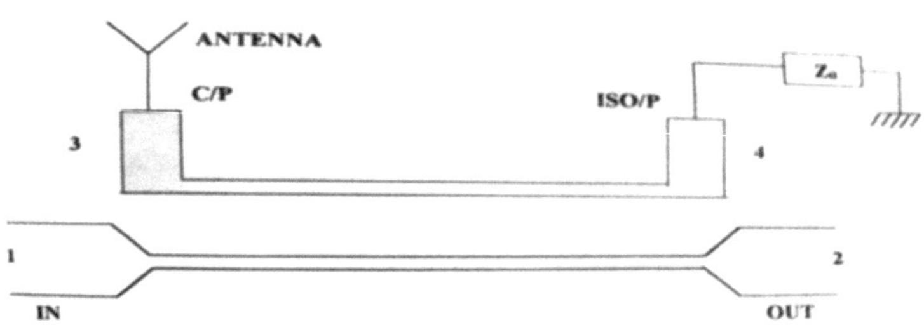

Figure 4.
Coupled line directional coupler.

2.7 90^0 Hybrid coupler lumped version

This coupler comprises of four quarter-wave sections, two in series and two in shunt. Each quarter line is realized by equivalent π section of lumped elements (inductors and capacitors) [2, 3].

In this structure, diagonally opposite ports are coupled. Since the coupled signal travels a distance of two quarter-wavelengths, it will be out of phase with respect to the input port. Powers given at the port-1 will be distributed between the direct output port-2 and coupled port-3. Port-4 is isolated. The coupling factor is dependent on the normalized impedance of series and shunt arm quarter-wave line sections, Z_b and Z_a respectively. Coupling factor is given by.

$$k = -Z_b/Z_a \tag{12}$$

Condition for impedance matching is given by.

$$1/{Z_b}^2 – 1/{Z_a}^2 = 1 \tag{13}$$

2.8 Excitation coefficients in the transmit mode

The normalized amplitudes (C_n) for all antenna elements with respect to the first antenna from the divider is given by

$$C_n = Vk_n \prod_{i=1}^{n-1} \sqrt{(1 - k_i^2)} \tag{14}$$

Where K_I is the coupling factor of coupler 'I' from the center (divider).

2.9 Excitation coefficients in the receive mode

The normalized amplitudes (C_n) for all antenna elements with respect to the first antenna from the divider is given by

$$C_n = k_n \prod_{i=1}^{n-1} \sqrt{(1 - k_i^2)} \tag{15}$$

Where K_I is the coupling factor of coupler 'I' from the center (divider).

2.10 Illumination efficiency

Since the weighing factors of the antenna elements are same for both transmit and receive modes, illumination efficiency will be same for both the modes. Illumination efficiency is defined as "the ratio of effective length of the sub-array to the physical length", which can be expressed as

$$\eta_{ill} = \frac{1}{16C_1} \sum_{n=1}^{16} C_n \tag{16}$$

which is found to be 79%.

Feeder network consists of fifteen couplers on either side of the divider/combiner to feed 16 antennas on either side. The coupler rating in decibels and the

Coupler No. (N)	Coupler Rating In db	Coupling Factor Value(K_n)
1,2	10	0.316
3,4	9	0.355
5,6,7	8	0.398
8,9	7	0.448
10,11	6	0.501
12,13	5	0.562
14	4	0.631
15	3	0.709

Table 2.
The coupler ratings and their corresponding coupling factor.

Sub Array (N)	Coupling Coefficients(Cn)	Sub Array (N)	Coupling Coefficients (Cn)
1	1.0000	9	0.7699
2	0.9488	10	0.7697
3	1.0112	11	0.6610
4	0.9454	12	0.6467
5	0.9908	13	0.5349
6	0.9090	14	0.4968
7	0.8339	15	0.4330
8	0.8611	16	0.4330

Table 3.
The coupling coefficients of the coupler.

corresponding coupling factors of all the fifteen couplers are given in **Table 2**. The corresponding coupling coefficients (C_n) are given in **Table 3**.

2.11 Feeder network efficiency

Feeder network efficiency is defined as "the ratio of power delivered to the sub-array by the feeder network to the power output of the transmitter, feeding the sub array'. I the receive mode it is defined as "the ratio of the power delivered to LNA by the feeder network to the total power developed at the terminals of all thirty-two antenna elements during reception". All feeder line components are assumed to the lossless in the computation of the feeder network efficiency. Normalized characteristic impedance of the system is assumed as unity in power calculations.

2.12 Transmit mode

When an input voltage of $\sqrt{2}$ volts is applied to the power divider, the voltage amplitude at the input of the first coupler will be one volt. The total power delivered to the sub-array is given by

$$P_{out} = 2\sum_{i=1}^{16} A_n^2 units \tag{17}$$

Where A_n is the input voltage fed from the coupled ports of the directional couplers. Power input to the feeder network is,

$$P_T = \left(\sqrt{2volts}\right)^2 = 2units$$

Feeder efficiency η_{ft} is given by is found to be 100%

$$\eta_{ft} = \frac{P_{ant}}{P_r} \tag{18}$$

2.13 Receive mode

In the receive mode, all the antennas of the sub-array equal powers and deliver the same to the coupled ports of the feeder network. When an input voltage of 1 volt is applied to all the coupled ports, the input power to the feeder network is given by

$$P_{ant} = 32 \times (1)^2 = 32 \text{ watts.}$$

The output voltage of the combiner (which is fed to the LNA) is:

$$V_{out} = \left(\sqrt{2}\right) V_{0,1} \tag{19}$$

Where, $V_{0,1}$ is the output of the first coupler. The combiner output power is given by.

$$P_{out} = 2|V_{0,1}|^2 \tag{20}$$

Feeder network efficiency is given by

$$\eta_{fr} = \frac{P_{out}}{P_{tn}} \tag{21}$$

and found to be 92.8%. Note that the rest of the power will be dissipated in the isolated ports of the directional couplers.

In addition to this, there will be a combining loss (at IF level) of 0.6 dB, which is equivalent to an efficiency of 92.3%, and due to amplitude imbalance there will be some loss that should be accounted in the overall feeder line efficiency. Hence, the total feeder efficiency of the MST radar planar phased array is equal to 85.6%.The specifications of MST radar are listed here:

Type of the array	Phased antenna array
No of elements	1024
Grid	Square
Configuration	32 × 32 matrix
Inter-element spacing	0.7 wavelengths
Physical aperture	130 m × 130 m

Aperture distribution		Stepped modified Taylor across the principal planes
Effective aperture		100 m × 100 m
Peak power aperture product		2.5×10^{10} W-m2
Gain		36 dB
Beam width		3° in the principal planes
Side lobe level		−20 dB
Beam steering		0 to 20^0 (in steps of 1°) from Zenith towards all directions
Feeder type		32 parallel runs of center-fed-Series-feed
Feeder efficiency (theoretical)	Transmit mode	100%
	Receive mode	92.8% in E-plane & 92.3% in H-plane
Feeder loss	Transmit mode	2.00 dB
	Receive mode	2.20 dB
Antenna element		Three-element Yagi-Uda
Frequency of operation		53 MHz
Wavelength		5.66 m
Length of the director		0.418 λ
Length of the exciter		0.471 λ
Length of the reflector		0.495 λ
Exciter-director spacing		0.158 λ
Exciter reflector spacing		0.219 λ
Radius of the element		0.00165 λ
Element pattern - Directive gain		7.8
Element pattern - Beam width (E-plane)		60^0
Element pattern - Beam width (H-plane)		88^0
Element pattern - Front-to-back ratio		14 dB

3. Aperture thinning of MST radar antenna Array

Most often, a few number of transmitters are non-operational due to various factors, making the linear sub-arrays corresponding to these transmitters ineffective. Even if the transmitters are operational, with in a sub-array, it is possible that some elements will not get the excitation signal due to weak connection or discontinuity problems in the feeder line. This results in the thinning of the aperture and the deviation of the excitation from the specified Taylor distribution. Due to this deviation, the array pattern will be distorted from the normal pattern. In this chapter, a detailed analysis is carried out to quantify the degradation in the radiation pattern due to aperture thinning. Phase-errors are assumed to the zero throughout.

3.1 Array pattern of 2 N MST radar

If the array aperture is in the x y-plane and sub-arrays are aligned parallel to y-axis with a spacing d_x, along the x-axis, array pattern can be expressed as

$$f(\theta,\phi) = \sum_{m=1}^{2Nx}\sum_{n=1}^{2Ny} I_{mn} e^{\left[-jkSin\theta\left(md_x Cos\phi + nd_y Sin\theta\right)\right]} \tag{22}$$

Where d_x = sub-array spacing along the x-axis
d_y = element spacing within a sub-array (along the y-axis)
m = sub-array number along the x-axis
n = element number along the y-axis within a sub-array
θ = Field point angle from broadside
Ø = Azimuth angle
I_{mn} = excitation current coefficient of nth element in m^{th} row (sub-array)
$2N_x$ = number of sub-arrays in x-axis
$2N_y$ = Number of elements within a sub-array (along y-axis)
k = Phase constant (in free space)

For MST radar antenna array dx = dy = d = 0.7λ and 2Nx = 2Ny = 2 N = 32. If each row has the same current distribution, even though the current levels are different in different rows, that is

$$\frac{I_{mn}}{I_{ml}} = \frac{I_{In}}{I_{11}} \tag{23}$$

Which is true for MST array case, and hence possible to separate the current distribution and the array factor can be expressed in the form

$$f(\theta,\varnothing) = f_x(\theta,\varnothing) f_y(\theta,\varnothing) \tag{24}$$

In which

$$f_x(\theta,\phi) = \sum_{m=1}^{2N} I_m e^{(-jmkd_x Sin\theta Cos\phi)} \tag{25}$$

$$f_y(\theta,\phi) = \sum_{n=1}^{2N} I_n e^{(jnkd_y Sin\theta Sin\phi)} \tag{26}$$

and

$$I_m = \frac{I_{m1}}{I_{11}} \tag{27}$$

$$I_n = \frac{I_{1n}}{I_{11}} \tag{28}$$

are the normalized current distributions in a row of elements parallel to x-axis and y-axis respectively. All the thirty-two elements are phase-equalized within a sub-array by adjusting the input feed cable lengths. Hence a linear sub-array, when excited alone, will produce a fan beam in the broadside direction. Beam tilting is done in E-plane (or Ø = 0° plane or xz-plane in this case) by providing progressive phase shift along the successive linear sub-arrays. The equations are executed in MATLAB to find the array patterns at different zenith angles.

If an array aperture is not fully excited, then it is said to be "Thinned". When this thinning is applied for the MST radar array, which is a planar array with separable current distribution, the array pattern can be expressed as

$$f(\theta,\phi) = \sum_{m=1}^{2N} I_m e^{(-jmkd_x Sin\theta Cos\phi)} \sum_{n=1}^{2N} I_n e^{(jnkd_y Sin\theta Sin\phi)} \tag{29}$$

where I_m is proportional to the square root of output power of transmitters and in is proportional to the coupling coefficients of CFSF network. The array factors in both the principle planes, Ø 0° (E-plane) and Ø = 90^0 (H-plane), respectively are

$$f_E(\theta) = f(\theta, 0) = \sum_{n=1}^{2N} I_n \sum_{m=1}^{2N} I_m e^{(-jmkd_x Sin\theta)} \tag{30}$$

$$f_H(\theta) = f(\theta, 90^0) = \sum_{m=1}^{2N} I_m \sum_{n=1}^{2N} I_n e^{(-jnkd_y Sin\theta)} \tag{31}$$

When some of the currents I_m are zero (which means the corresponding transmitters are off), it is clear that the shape of H-plane pattern, $f_H(\theta)$, will not be affected though its magnitude changes according to the first term in (31). However, the E-plane pattern, $f_E(\theta)$, will be distorted according to the second term of (30). So, when few transmitters are non-operational only the e-plane pattern will be distorted. To study the effect of aperture thinning on the radiation pattern, it is required to compute in MATLAB the array pattern by letting some of the I_m to zero, which is equivalent to putting the corresponding transmitters off.

4. Results and discussions

The antennas in the sub-arrays would not get excitation signal, if few transmitters are non-operational. Though they are physical present, electrically they are not effective. If an array aperture is not fully excited, then it is said to be 'thinned'. This results in the deviation of the excitation from the specified Taylor distribution. Due to this deviation, they array pattern will be distorted. To quantify the distortion in the radiation patterns in both E and H planes, the array pattern expressions are made to depend on the shape of the beam. Forcing some of the I_m to zero, which means these transmitters are non-operational, can effect array thinning. Iy is clear that the shape of H-plane pattern, $f_H(\theta)$, will not be affected, though its magnitude changes according to the first term in Eq. (31). However, the E-plane pattern, $f_E(\theta)$, will be distorted according to the second term of Eq. (30) Hench, we can conclude that only E-plane pattern will be distorted and needs to the examined in case few transmitters are non-operational. Programming in MATLAB helped a lot to examine these cases.

Array pattern is computed in the E-plane with different thinning configurations, that is by letting group of transmitters (or sub-arrays) in effective. Radiation parameters are distorted for all the cases. The two important parameters that may affect the radar performance are gain and SLL. The variation of these two parameters with different array thinning configuration is tabulated in **Table 4**. Array pattern thinning obtained with and without tilting can be viewed from plot shown in the **Figures 5** and **6**.

Array pattern is computed for different azimuth angles using Eq. (30) by letting all transmitters (or sub-arrays) effective. The important parameters that may affect the radar performance are gain and SLL. The variation of the SLL parameter is tabulated in **Table 5**. Array pattern computed for different azimuth angles obtained without thinning and tilting can be viewed from **Figures 7–9**.

The antenna array 3-D array pattern obtained using MATLAB is shown in the **Figure 10**. Amplitude distribution is plotted using the Eq. (22). **Figure 11** shows a 3–D plot of antenna array when fully excited. Distortions of amplitude distribution due to array thinning are shown in **Figures 12** and **13**.

S.NO	Sub-arrays or TXs OFF	Loss in the gain (dB)	SLL (dB)
1	Nil	0.00	−19.9
2	3,4	0.40	−19.9
3	5,6	0.50	−22.0
4	3,4,5,6	1.00	−19.5
5	5,6,27,28	0.55	−17.5
6	1 to 8	2.00	−16.0
7	1 to 8 and 25 to 32	4.00	−13.5
8	9,10	0.50	−18.3
9	11,12	0.80	−17.7
10	13,14	0.80	−15.2
11	15,16	0.80	−14.2
12	13,14,15,16	1.90	−11.5
13	15,16,17,18,19	2.00	−09.0
14	9 to 6	3.30	−07.0
15	6,8,9,17,24,29,30,32	2.50	−14.5

Table 4.
Variation of parameters with different array thinning configurations.

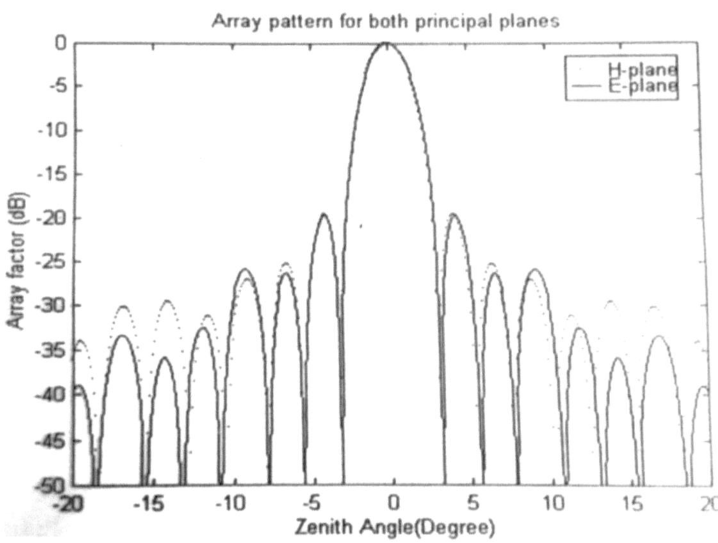

Figure 5.
Array pattern for the two principal planes.

Polar plots plotted in MATLAB for both the principal planes are shown in the **Figures 14** and **15**. Radiation pattern for different azimuth angles obtained in MATLAB are shown in **Figures 16** and **17**.

4.1 MATLAB package

The entire software is developed using the MATLAB package. MATLAB was chosen for the numeric computation and visualization of the array pattern. Matalb

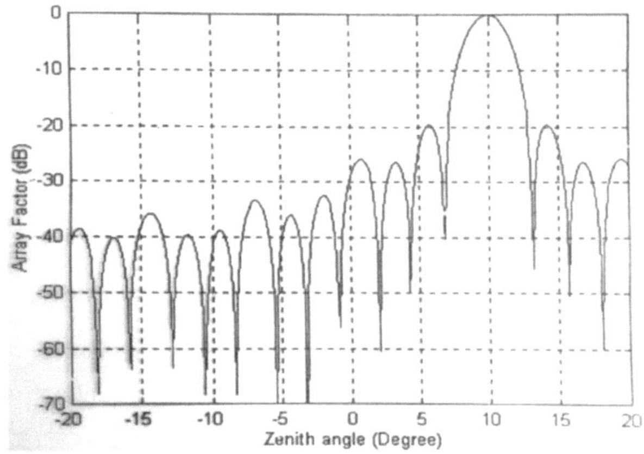

Figure 6.
Array pattern for a tilt of 10 degrees in E-plane.

S.No.	Azimuth angle (Ø)	SLL (dB)
1	0	−19.9
2	15	−21.5
3	30	−30.5
4	45	−39.5
5	60	−30.5
6	75	−21.5
7	90	−19.9

Table 5.
Variation of SLL standard configuration with different azimuth angles.

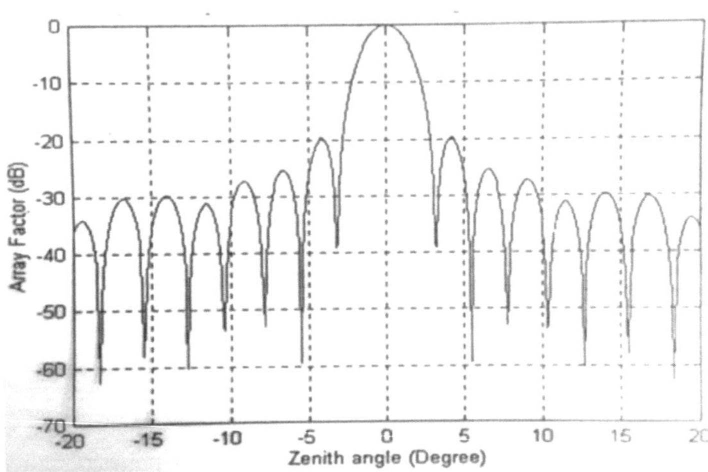

Figure 7.
H -plane pattern – without distortion and with a tilt of 10 deg.

is a helpful tool to develop portable and graphical user interface software. After entering into the package, the menu options as shown in **Figure 18** will be visible. The new file with the rated powers of the transmitters, and standard

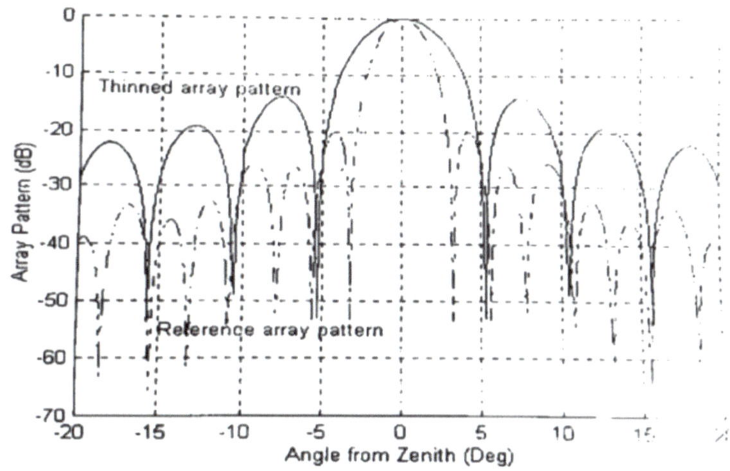

Figure 8.
E plane pattern distortion due to array thinning transmitters off: 1 to 8 & 25–32 and tilt: 0 deg.

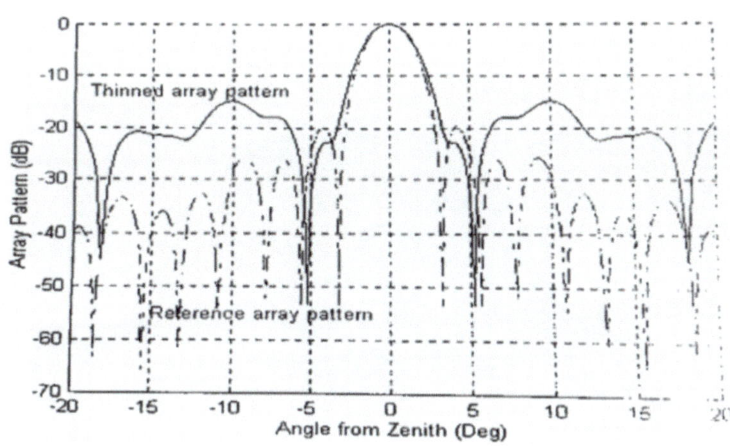

Figure 9.
E- plane pattern distortion due to array thinning transmitters off: 6, 8, 9, 17, 24, 29, 30 & 32 and tilt: 0 deg.

Figure 10.
3-D Array pattern.

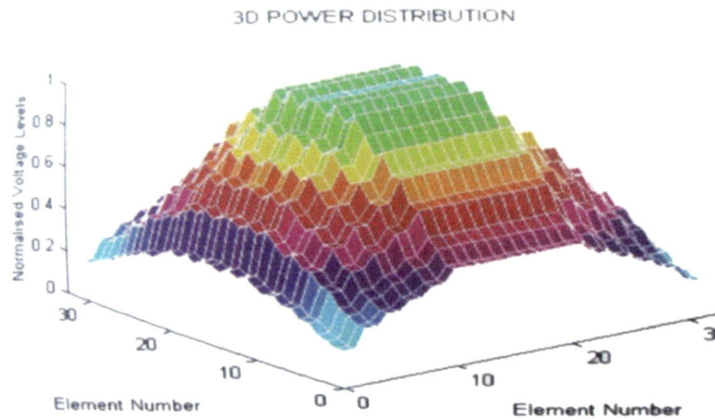

Figure 11.
3-D power distribution for MST radar antenna array.

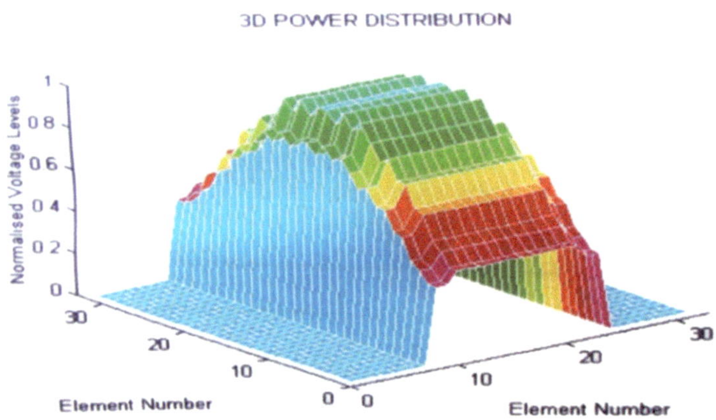

Figure 12.
3-D power distribution for MST radar antenna array transmitters off: 1–8, & 25–32.

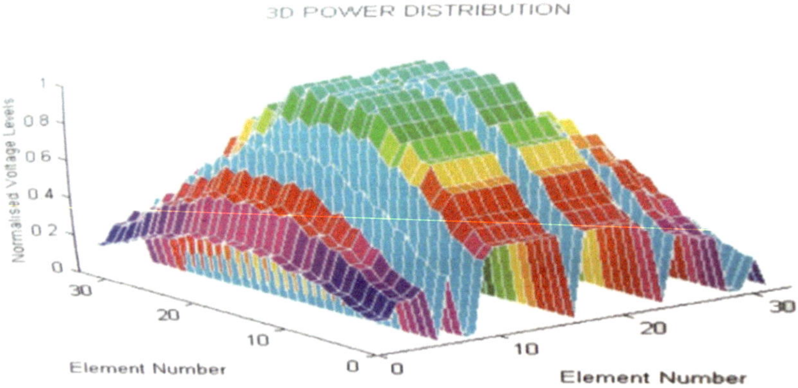

Figure 13.
Transmitters off: 6, 8, 9, 17, 24, 29, 30 & 32.

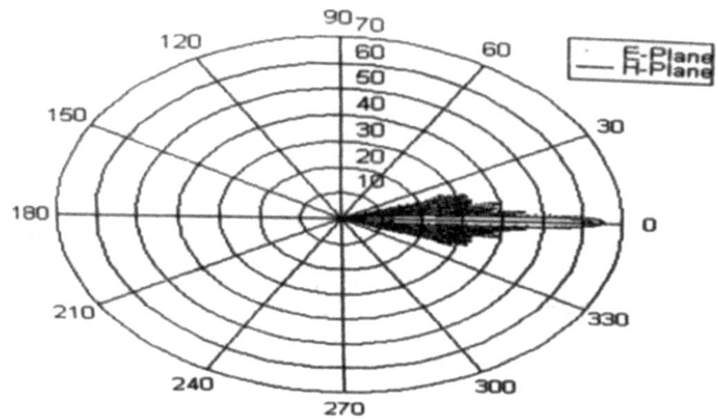

Figure 14.
Radiation power pattern for both principal planes.

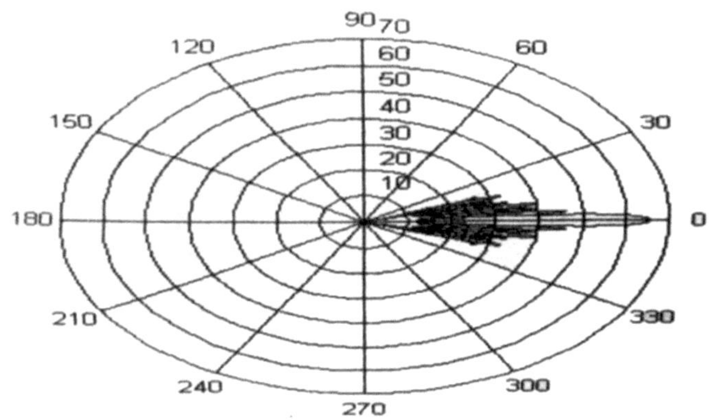

Figure 15.
Radiation power pattern for E-plane.

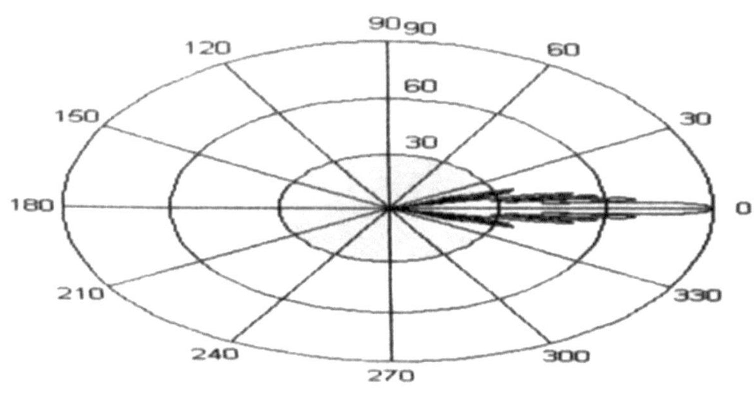

Figure 16.
Radiation power pattern for an azimuth angle of 15°.

parameters of the MST radar can be selected and will have. dat extension.
Figure 19 shows the window of the open file with open options such as m files, mat files etc.

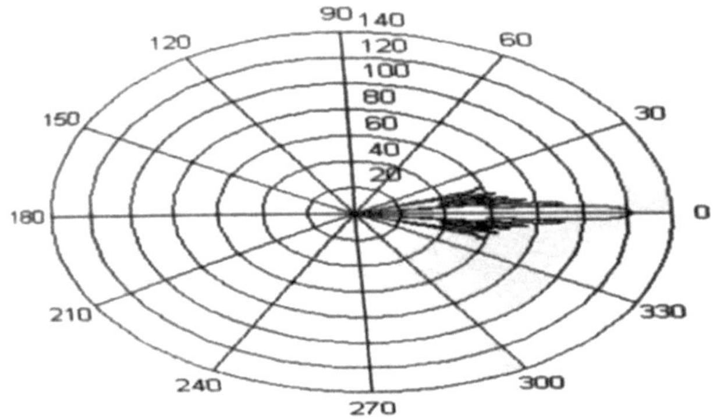

Figure 17.
Radiation power pattern for an azimuth angle of 30°.

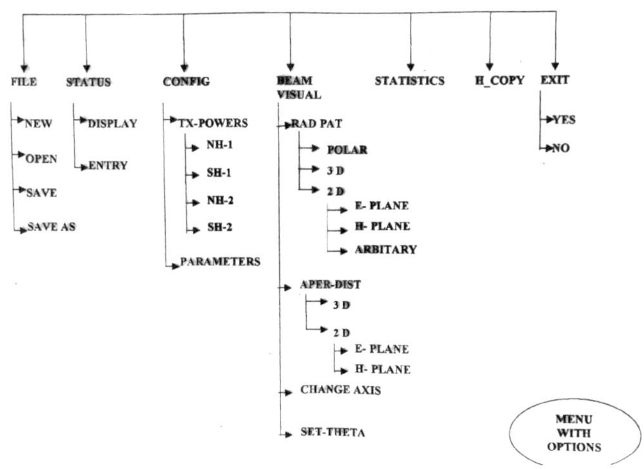

Figure 18.
Array factor distortion analysis menu options.

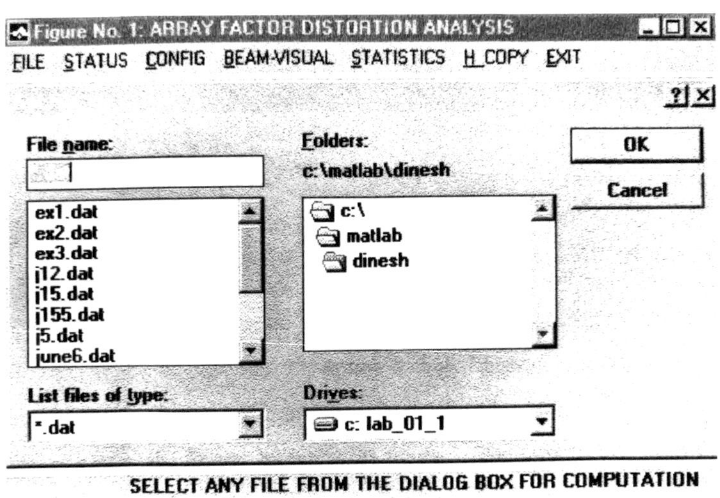

Figure 19.
Array factor distortion analysis file open selection option.

DOI: http://dx.doi.org/10.5772/intechopen.97637

The display window will be of the format shown in **Figure 20**. The format of the display window is of 6 x 8 matrix. Press <O.K > button to clear the display. The sub menu allows the user to view as well as enter the values of your choice for both transmitters and parameters values, for the further processing. Format for entering the values in 6 x 8 matrix is shown in **Figure 21**. The accept button is pressed for processing the entered values.

The Config option allows the user with two sub menu options 1) Tx-power and 2) Parameters. The TX-Power option allows the user to change the transmitter power levels with the help of the slider as shown in **Figure 22**. The TX-Power have 4 sub menu options for each hut that is NH-1, NH-2, SH-1, and SH-2. The user can

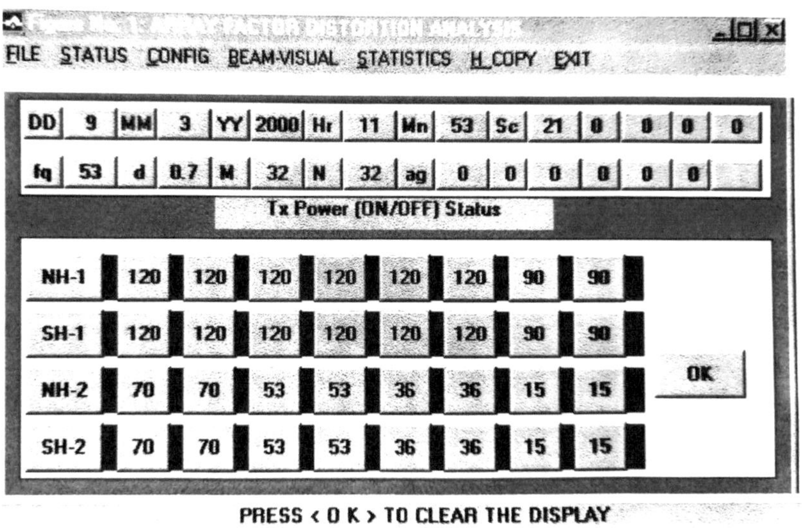

Figure 20.
Transmitter power ON/OFF status.

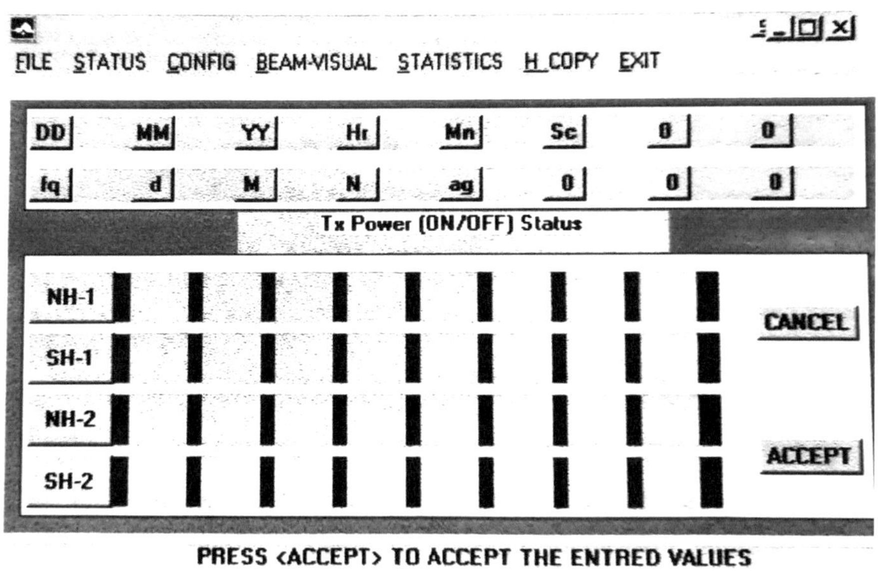

Figure 21.
Format for entering transmitter power values.

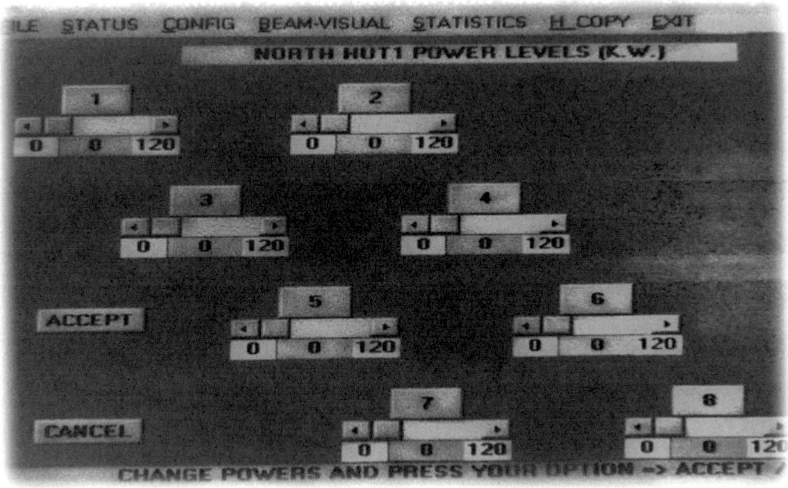

Figure 22.
The transmitter power levels changes in north hut 1 (NH 1).

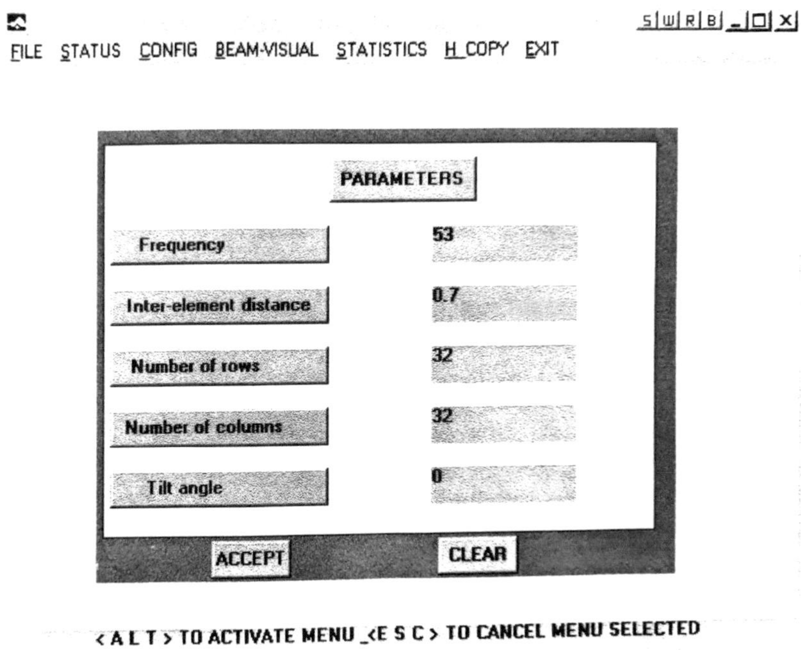

Figure 23.
Option to change the antenna array parameters.

change the parameters of the antenna array as shown in **Figure 23**, where the operational frequency in MHz, inter-element distance in wavelengths, number of rows and columns, tilt angle of the beam in degrees can be varied.

5. Conclusions

From the results obtained in MATLAB, it may be noticed that degradation in gain and SLL due to the absence of a few low power transmitters (**Table 4**)

(1–4 cases) is not significant. Surprisingly, SLL improves with some low power transmitters off (case-3), where the loss in directive gain is only marginal symmetrical thinning gives higher SLL than asymmetrical thinning (cases 4 & 5). If all low power transmitters are off (case-7), then the array is similar to standard radar array (almost uniform distribution) giving a SLL of −13.5 dB.

On the contrary, the absence of high power transmitters causes SLL to increase significantly. From cases 8–11 of the table, it is clear that absence of even two transmitters increases the SLL by 1.5–6 dB. It may be noted that SLL depends on the position of transmitters that are off. Absence of central sub-arrays results in higher SLL. Cases 12–14 demonstrates that absence of more than four power transmitters give unacceptable SLL (worse than Uniform distribution case). Finally, case 15 represents the real time status of the radar on a particular operational day, where eight transmitters were not functioning.

For different azimuth angles, Radiation power pattern of the MST radar antenna array is plotted. **Table 5** shows the radiation power pattern for different azimuth angles. and the variations of SLL for standard configuration of radar array, with different azimuth values. It is observed that the change in the SLL for different azimuth angles varies a lot compared to that of the SLL of the two principal planes. The main focus is on the analysis of distortion of array pattern due to thinning.

The following facts can be concluded from the pattern obtained using MATLAB:

- The first SLL rises by 0.4 dB.
- For a continuous distribution SLL taper off very fast. Far off, side lobe levels do not taper off fast.
- Around 45^c the side lobe level rises roughly by 10 dB
- 3 dB beam width [15] found to be same.

The work presented in this report can further be extended to study in the following cases.

- Element pattern integrating with group pattern using MATLAB Antenna toolbox
- Mutual Coupling between antennas in the MST Radar antenna array using MATLAB Antenna toolbox

Acknowledgements

I deem it as privilege to acknowledge my indebtedness to all those people who have helped me in completing the investigation of the. I express my sense of gratitude and thanks to my guides Dr. P Srinivasulu, Engineer ‘SG’, NMST Radar Facility, Gadanki and Dr. N C Eswar Reddy, Professor, Sri Venkateswara University, Tirupathi for their valuable guidance, encouragement, inspiration and cooperation throughout the investigation.

Author details

Nali Dinesh Kumar
Vignan Institute of Technology and Science, Affiliated to JNTUH, Hyderabad, India

*Address all correspondence to: nalidinesh@gmail.com

References

[1] MST Radar manual, "Characterization of MST Radar Antenna Array", P. Srinivasulu, Scientist-F.

[2] Ricolindo L. Carino and Mark.F. Horstemeyer (July 7th 2016). Case Studies in Using MATLAB to Build Model Calibration Tools for Multiscale Modeling, Applications from Engineering with MATLAB Concepts, Jan Valdman, IntechOpen, DOI: 10.5772/62348.

[3] Bahadır Ergün and Cumhur Şahin (January 6th 2021). Laser Point Cloud Segmentation in MATLAB [Online First], IntechOpen, DOI: 10.5772/intechopen.95249

[4] Constantine A. Balanis, "Antenna Theory analysis and Design", (2nd edition). John Wiley & sons Inc. New York, 2005.

[5] Edward C. Jordon and Keith G. Balman, "Electromagnetic Waves and Radiating Systems", (2nd edition), Prentice Hall of Indian Pvt. Ltd., New Delhi, 1991.

[6] N Dinesh Kumar, "Array Factor Distortion Analysis", Verlag Publisher: LAP publishers, 1st edition.

[7] H. Steyskal, "Simple method for pattern nulling by phase perturbation", IEEE Transaction on Antennas and Propagation, vol.31, pp.163-166, 1983. 10.1109/TAP.1983.1142994

[8] V.Rajya Lakshmi and G.S.N.Raju, "Optimization of radiation patterns of array antennas", PIERS Proceedings, Suzhou, China, pp.1434-1438, 12-16 September 2011

[9] Keen-Keong Yan and Yilong Lu, "Sidelobe reduction in array-pattern synthesis using genetic algorithm", IEEE Transactions on Antennas and Propagation, vol.45, no.7, July 1997. 10.1109/8.596902

[10] Haupt, R. L. and Y. Rahmat-Samii, "Antenna array developments: A perspective on the past, present and future," IEEE Antennas and Propagation Magazine, Vol. 57, No. 1, 86–96, 2015. doi:10.1109/MAP.2015.2397154

[11] Farina, A. and L. Timmoneri, "Phased array systems for air, land and naval defence applications in Selex ES," 8th European Conference on Antennas and Propagation (EuCAP), 560–564, Hague, 2014. doi:10.1109/EuCAP.2014.6901818

[12] Ruze, J., "The effect of aperture errors on the antenna radiation pattern," Nuovo Cimento Suppl, Vol. 9, No. 3, 364–380, 1952. 10.1007/BF02903409

[13] Wang, C., M. Kang, W. Wang, B. Duan, L. Lin, and L. Ping, "On the performance of array antennas with mechanical distortion errors considering element numbers," International Journal of Electronics, Vol. 104, No. 3, 462–484, 2017. doi:10.1080/00207217.2016.1218064

[14] Wang, C., et al., "Electromechanical coupling based performance evaluation of distorted phased array antennas with random position errors," International Journal of Applied Electromagnetics and Mechanics, Vol. 51, No. 3, 285-295, 2016. doi:10.3233/JAE-150170

[15] Hsiao, J. K., "Array sidelobes, error tolerance, gain and beamwidth," Naval Research Lab Report, 8841, Washington DC. Sep. 28, 1984.

Section 3

Geometric Segmentation

Chapter 3

Laser Point Cloud Segmentation in MATLAB

Bahadır Ergün and Cumhur Şahin

Abstract

Currently, as a result of the massive continuous advancements in laser measurement technology, possibilities of map production are broadened, the loss of time and the waste of material sources are highly prevented, and the accuracy and precision of the obtained results are significantly improved. In the view of engineering concept. However, big data which are from laser point clouds have been especially used in the significant procedures of surveying studies. Programming methods are dependent in each studies. In the necessity of the applications, the coding procedure has more efficient, the data of work has increased, and time has been consumed. The coding methods have necessarily been optimized for working together especially in the big data studies. In this section, an automated survey (building facade surveying) is produced from scanning data by means of coding in MatLAB.

Keywords: surveying, laser point cloud data, segmentation, object determination, coding in MatLAB

1. Introduction

Nowadays, laser scanning and modeling technology have been extensively used in city documentation and cultural heritage studies besides the technique of imaging for global representation in the internet survey and navigation applications. In the view of engineering concept. However, big data which are from laser point clouds have been especially used in the significant procedures of surveying studies. Programming methods are dependent in each studies. In the necessity of the applications, the coding procedure has more efficient, the data of work has increased, and time has been consumed. The coding process of big data process must be modeled in the hardware systems requires receiving consideration. This process has been related to hardware construction by electronic and computer engineering vision. This process has mathematical model and algorithm, which has been concerned surveying and computer programming engineering vision. In this chapter, Matlab coding models including functional and stochastic properties have been suggested and discussed for operational process within laser scanning data segmentation in surveying studies.

2. Process

2.1 Data structure

Laser scanners scan the object in horizontal and vertical directions under a certain angle as a series of points and this allows the object to be displayed as a point cloud. In order to determine each of the laser points, measurement of scanner-centric polar coordinates are made [1]. The measured points are slanted distance to point P, the angle between X-axis and horizontal plane α, and the angle of inclination of the horizontal plane measuring line φ. As illustrated in **Figure 1**, the initial point of terrestrial laser scanners are considered to be the positioned points. These measurements are based entirely on their local coordinate systems.

The resulting point cloud data is processed in formats related to the coordinate and the angle. The processing is carried out as follows: DXF for CAD models, ASCII for surface modeling, VRML format for visualization, and txt or pts. Software which varies with different laser scanner instrumentation can be used to obtain the point cloud data.

Laser scanners initially obtain the X,Y,Z Cartesian coordinates inside a second coordinate system which is located at the center of the station point and then they scan the surface of the object. In addition to the three-dimensional coordinates, the resulting data includes the density of the returning signal in terms of RGB (Red, Green, Blue) depending on the structure of the surface in question and the distance of measurement. Modeling of the scanned object and environment gets easier with recorded RGB density values. The dense data obtained by scanning is called a point cloud. **Table 1** displays the formation of txt data linked to the point cloud data.

There is a software in target-oriented modules to obtain raw data, to convert data to a workable format, and to perform the texturing process (if necessary) etc.

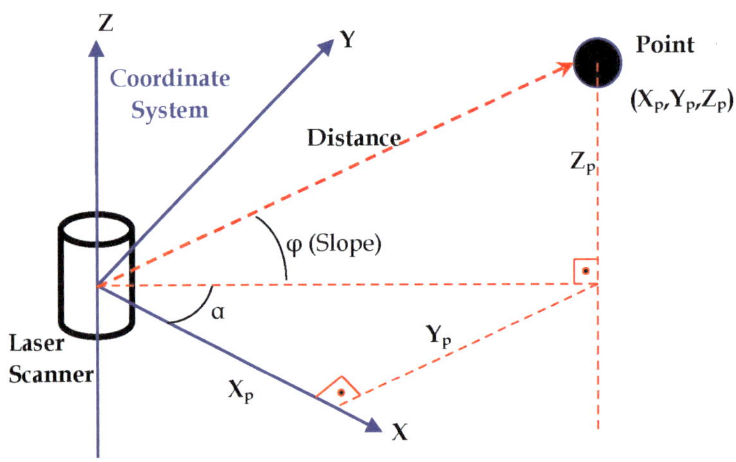

Figure 1.
TLS local coordinate system.

Geometric Information			Radiometric Information			
X (m)	Y (m)	Z (m)	Gamma	Red	Green	Blue
0.153229	0.521369	−0.004161	−76	91	115	113
0.270996	0.521319	−0.004880	−75	87	109	107
0.153229	0.467538	−0.009394	41	75	94	98
0.270874	0.467157	−0.006026	−167	81	100	97
0.216461	0.494419	−0.006889	−170	79	98	96

Table 1.
Point cloud data formation in ASCII format.

Figure 2 shows the point cloud data that is represented with RGB density values. A Leica HDS – 3000 terrestrial laser scanner is used to scan the point cloud data.

2.2 Programming flow procedure

During the documentation of the coordinate information of laser scanner point cloud data, there is no regular data order and classification [2]. For segmentation, points with known three-dimensional coordinates must be selected from all point cloud data. The algorithm is formalized with mathematical surface or point clustering techniques. Planar surfaces or points including depth parameters can be extracted by using various methods [3–9]. In addition, the surface points of the assigned surfaces are filtered. Once the building's planar surfaces are obtained, various methods can be used to extract property boundaries.

Programming flow procedure could be designed in this method in Matlab programming in **Figure 3**.

Deciding segmentation method is a dynamic process in Matlab programming flow. There are three different segmentation methods have been used in this step.

Geometric segmentation is based on geometric information of point cloud data. Radiometric segmentation is based on radiometric information of point cloud data. Hybrid segmentation is based on all information of point cloud data. The

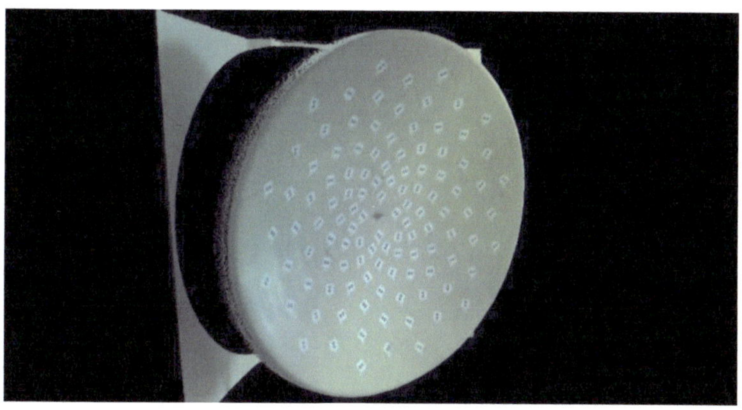

Figure 2.
Cyclone 5.2 point cloud image.

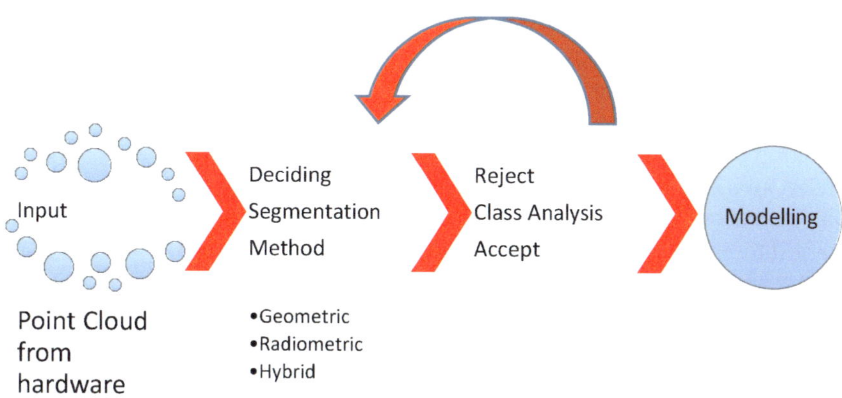

Figure 3.
Programming flow procedure.

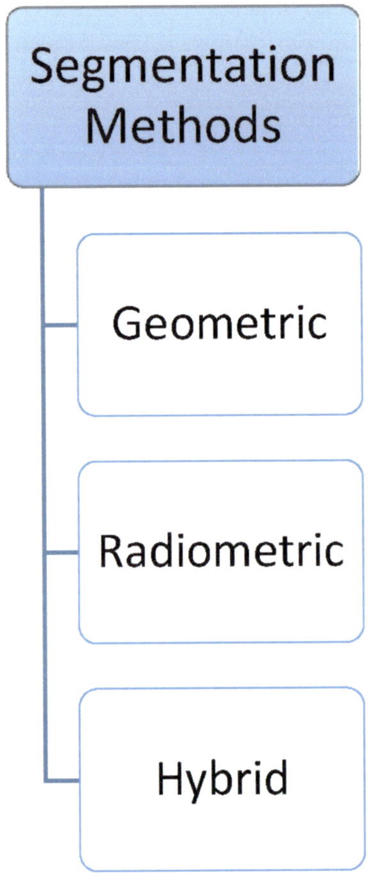

Figure 4.
Segmentation methods.

mathematical model of these segmentation methods could be based on not only conventional methods but also expert systems (Fuzzy systems, SVM, etc.) as shown in **Figure 4**.

3. Examples of geometric segmentation in Matlab

3.1 Point segmentation

Algorithm of this study which aims at filtering laser point cloud data of parallel surfaces in indoor areas by the help of filtering function is shown in **Figure 5** [10].

Selecting one of the parallel indoor surfaces as the reference plane by an operator is the first step of filtering function algorithm.

Distinct surfaces define the point cloud data which is illustrated in **Figure 6**. Planar surfaces like walls generally define the indoor areas. **Figure 6** displays point cloud data and various surface structures in a three dimensional coordinate system.

Mathematical function that represents plane surface is given in Eq. (1).

$$Ax + By + Cz + D = 0 \tag{1}$$

Eq. (1) shows that surface function consists of 4 parameters: A, B, C, D. The selected reference plane surface can be expressed mathematically by calculating

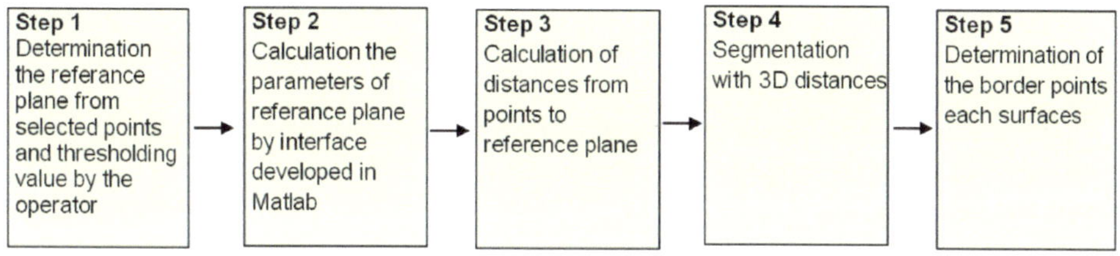

Figure 5.
Geometric segmentation steps.

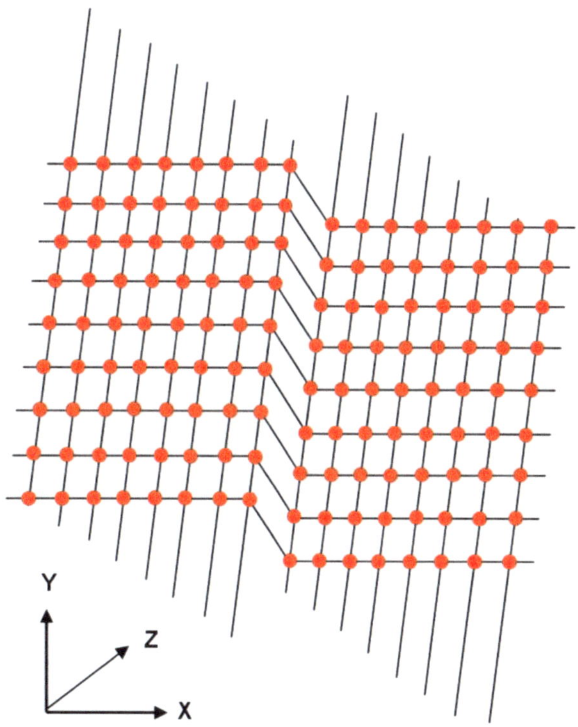

Figure 6.
Point cloud data structure in 3D coordinate system.

these four parameters. An operator must read and manually enter the selected point coordinates of reference plane. The parameters of selected reference plane surface are determined by this way. A plane can be mathematically defined with reference to four parameters. So, we can only define the parameters of reference plane with four points. An operator selects more than four points in the same reference plane and determines the parameters of reference plane in adjustment process. Operator manually enters the threshold value secondly. Threshold value can be defined as the minimum difference of depth during the filter operation. All the operations are performed automatically by a Matlab-based software, other than the two stages of the algorithm.

Calculating the parameters of the selected reference plane is the second step of geometric segmentation algorithm. Once the operator manually chooses multiple (five or more) points as part of the first step of algorithm, the Matlab-based interface automatically determines four parameters which represent the reference plane in the adjustment computation. Thus, the adjusted reference plane is created in this step.

The third step of the algorithm is to calculate the distances of the parameters specified in the adjustment computation and all points in the laser point cloud to the reference plane.

Figure 7 illustrates a point's distance to a plane. The distances of all points of the laser point cloud to the adjusted reference plane is calculated with Eq. (2). Eq. (2) shows the filtering function for the segmentation.

$$h = \frac{|Ax_1 + By_1 + Cz_1 + D|}{\sqrt{A^2 + B^2 + C^2}} \tag{2}$$

where;
h: Distance of a laser point to the reference plane,
A, B, C, D: Parameters of the adjusted reference plane,
x_1, y_1, z_1: Each laser point's three dimensional coordinates.

These stated distances are added in vector form to segmentation matrix S as a column. For each point in segmentation filtering, point distances for the reference plane are calculated separately. The definition of Matlab operation environment is done in the format of segmentation matrix. This is basically called the "segmentation matrix". However, column algorithm are used to make the calculations and final column vector provides classification of points on the surfaces. The first column of segmentation matrix which is shown in **Figure 8** is the X value, and the second column is the Y value of points. The third column represents the Z value in terrestrial laser scanning point cloud data. The fourth column shows the point distance into the reference plane. The fifth column displays the statistical

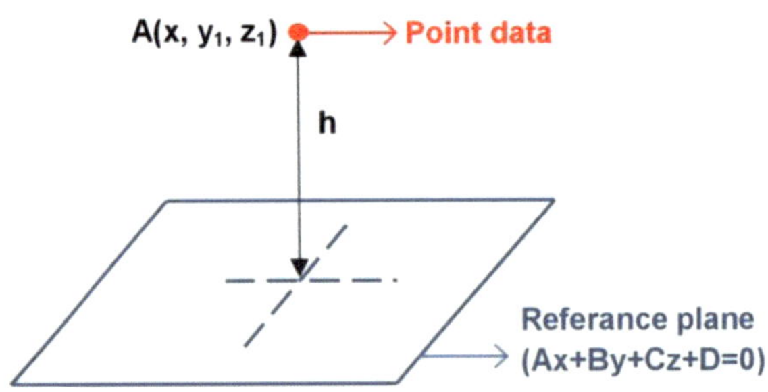

Figure 7.
Distance of a point to a plane.

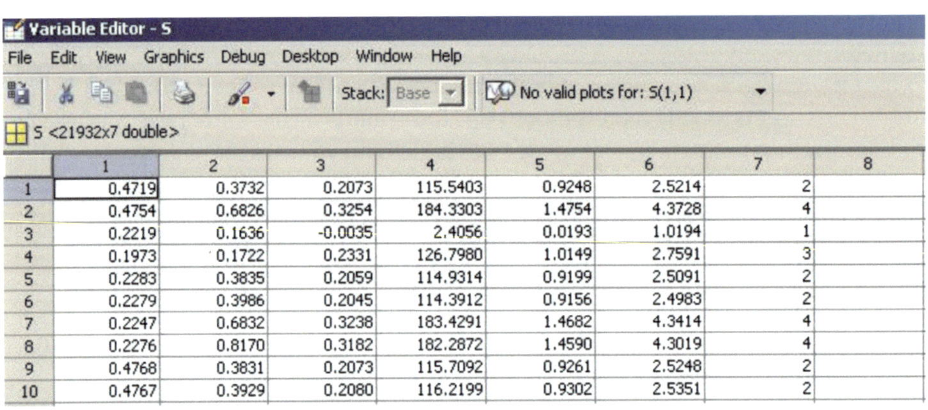

	1	2	3	4	5	6	7	8
1	0.4719	0.3732	0.2073	115.5403	0.9248	2.5214	2	
2	0.4754	0.6826	0.3254	184.3303	1.4754	4.3728	4	
3	0.2219	0.1636	-0.0035	2.4056	0.0193	1.0194	1	
4	0.1973	0.1722	0.2331	126.7980	1.0149	2.7591	3	
5	0.2283	0.3835	0.2059	114.9314	0.9199	2.5091	2	
6	0.2279	0.3986	0.2045	114.3912	0.9156	2.4983	2	
7	0.2247	0.6832	0.3238	183.4291	1.4682	4.3414	4	
8	0.2276	0.8170	0.3182	182.2872	1.4590	4.3019	4	
9	0.4768	0.3831	0.2073	115.7092	0.9261	2.5248	2	
10	0.4767	0.3929	0.2080	116.2199	0.9302	2.5351	2	

Figure 8.
Matrix form of segmentation algorithm in Matlab software.

differences in terms of distances. The sixth column illustrates the exponential values of the differences, and the seventh column shows the surface of the related points.

Algorithm's fourth step is called the step of segmentation (classification). This step is utilized for statistical vectorial change which is created for the study. In this step, the detection of the number of surfaces in a vector is essential and this is determined by the amount of statistical deviation. The values of standard deviation determine the statistical analysis. Standard deviation's comparison value is called the threshold value. Thus, the minimum difference of depth in question determines the amount of the statistical deviation. So, if less minimum difference of depths are obtained for the points, it means that they are on the same surface. In statistical analysis (5th Column) the total number of various surfaces is detected, and all distances are calculated as an exponential function (6th Column) which are shifted into a positive value. That is, the raw data is obtained to carry out the classification step. The exponential part of the obtained value is taken because there are some conditions. These conditions include a negative plane point distance and a stable point on the other side of the plane. So, the surface with a smaller absolute value gets the points between two surfaces. With reference to this value, each point in the laser are assigned to a surface and to the matrix of the surface (7th Column).

In the final step of the segmentation algorithm, four boundary surface points are defined by laser scanning points assigned to the surface matrix. For this purpose, minimum and maximum x, y plane coordinate values of the points in the surface matrix are used. The edges in each segment of the original segmented laser point cloud data are determined with reference to minX-minY, minX-maxY, maxY-minX, maxX-maxY values. So, the boundary points of both surface and x and y plane coordinates are evaluated. However, the original values are not considered as the height value of points with plane coordinates defined in the algorithm. If we created a surface with a Z value acquired from the laser scanning data of the four points of the surface, not every surface would be parallel to each other. Thus, if we assign a height value to the four edge points, the segmentation result is assigned as the average height value of the Z (height) value for all points assigned to that surface. Therefore, the Z value of the four segmented points for each surface is the average Z value. This is the same value for each of the four points. This can be seen in **Figure 9** [10].

At the end of the segmentation process, each surface with thousands of points is converted to planes that contain only four points and the average height of all points in a segment are taken. In order to test the filtering algorithm, laser point cloud of a

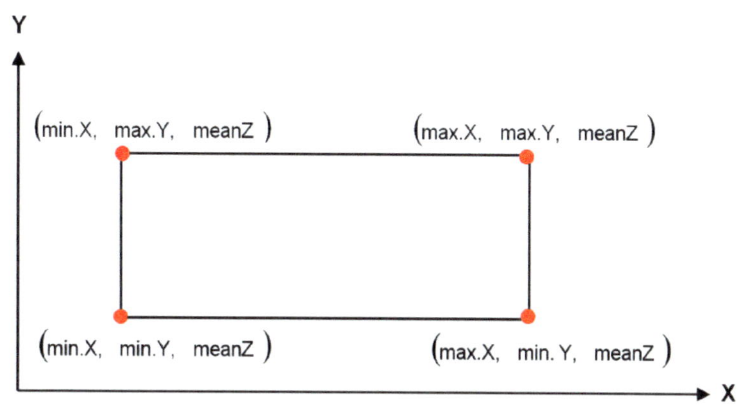

Figure 9.
Four corners of the surfaces obtained by segmentation.

class are scanned with Leica HDS 3000 in GYTE (presented in **Figure 10**). The scanning frequency in this study is 5 mm.

When obtaining the data which are specified in txt format from CYCLONE (Leica) software, affine conversion is carried out in order to give the data depth Z for point cloud data. So, operator will be able to determine the reference surface easier. An original data set, which is composed of four different surfaces and 21,932 points, is selected. There are significant depth differences which ranges from 1 cm to 20 cm in terms of the surfaces of doors, borders, walls and columns.

When the indoor space in **Figure 11** is examined, it is seen that surface 1 is the reference plane which enables the testing of the filtering function from four plane

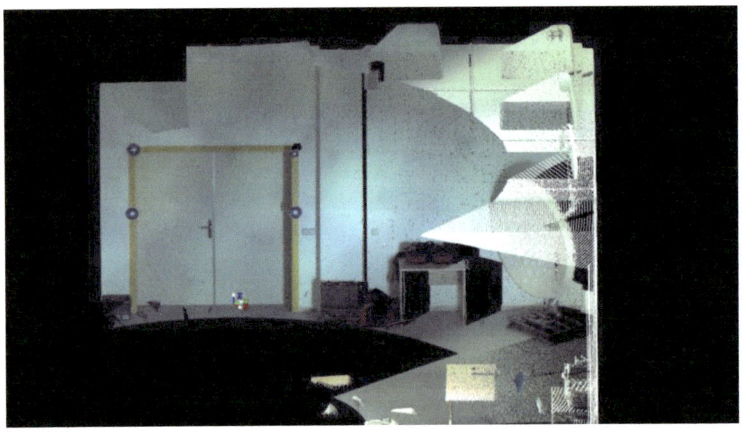

Figure 10.
Application site.

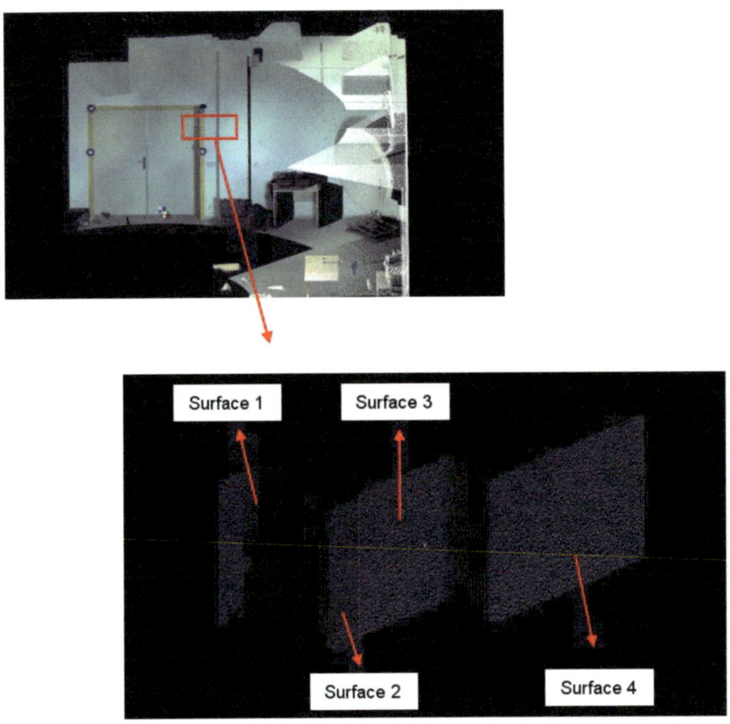

Figure 11.
Application data. (four indoor plane surfaces with different depths).

surfaces that are parallel to each other. **Figure** 7 shows that surface 1 is the backmost and surface 4 is the foremost of the surfaces. Operator chooses five points on the reference surface in Cyclone screen in order to calculate equation parameters of the plane. Threshold value (minimum difference of depth) of test data is 1 cm. By using the Matlab-based software, the classification of 21,932 three dimensional laser point clouds are made with four separate surfaces. This classification can be seen in **Figure 12**.

Figure 13 shows the graph which presents the exponential values. There are 3906 points in the first segment which is based on 21,932 original laser scanning data with four distinct classes. In the second segment, there are 6588. In the third segment, there are 1951 points, and in the fourth 9487 points (5 points belong to surface1). **Figure 14** presents the surfaces for each of the laser point cloud data. These are classified under four classes. **Figure 14** shows that the distance threshold value between the second and third plane surfaces is nearly 1 cm. The filter performs the segmentation of these two distinct surfaces without any errors. While assigning all the points on the laser point cloud into a plane surface, it is important to record them to their respective surface matrix. The points within the matrix of each surface enable us to find the X, Y plane coordinate values of four edge points of the plane surface. In this plane surface, only one Z value is designated to all four points.

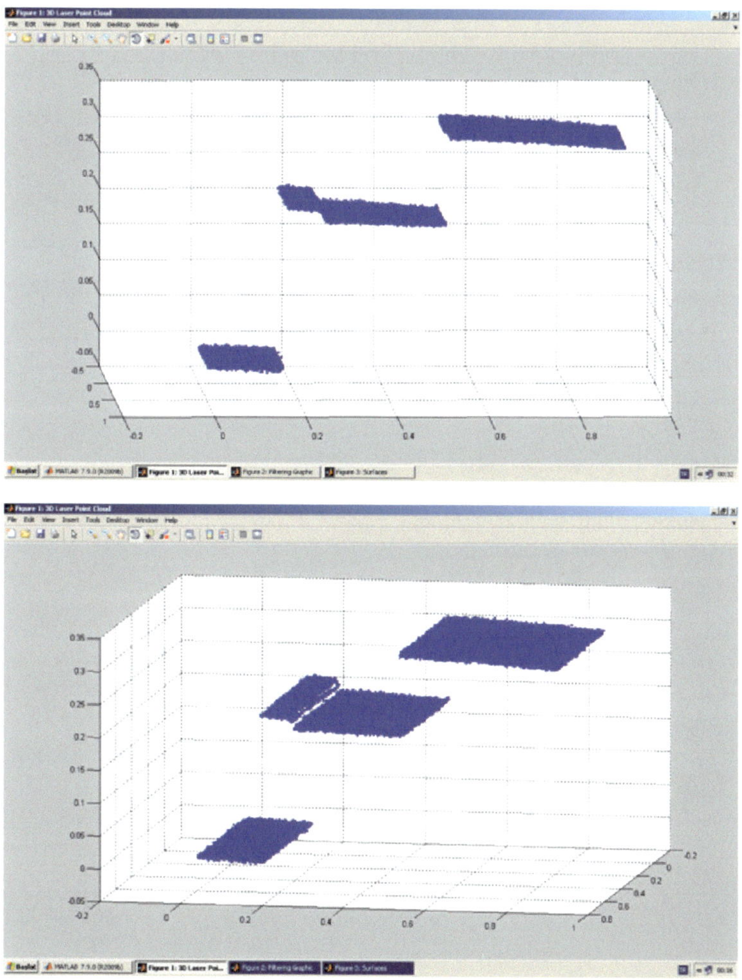

Figure 12.
Application data in Matlab software.

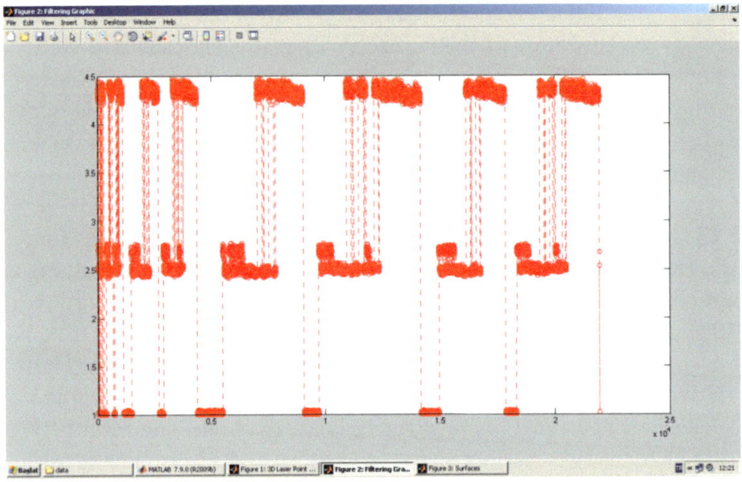

Figure 13.
Segmantation clustering graphic.

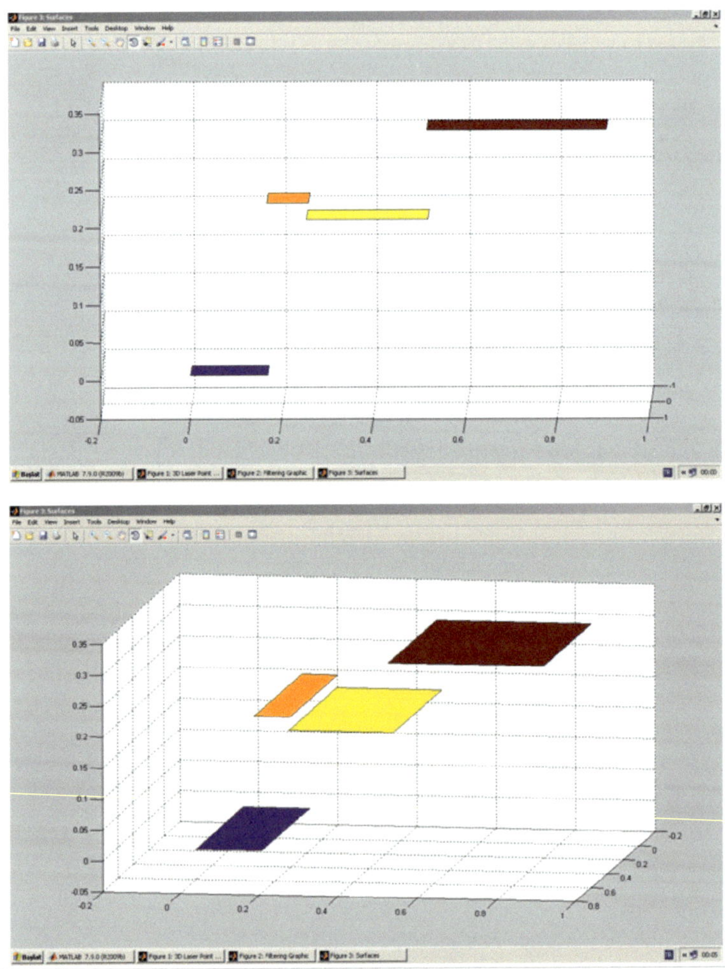

Figure 14.
The result of segmentation (four segmentated surfaces which are different depth).

The crucial consideration here is to understand that surface selection does not cause any limitation and the operator decides this surface and defines it as the reference plane. Operator can select any surface as the reference plane in the suggested algorithm. The closest or the farthest plane surface or any surface within these surfaces can be determined as the reference surface by the operator. So, exponential values are used during the process of creating the segmentation matrix (matrix S). Distances between points and reference plane can be either negative or positive but since the exponential values are used, this does not change the results of segmentation or classification. Thus, negative or positive distances to the reference plane do not result in any segmentation error (this only shows the side that they belong to on any plane and does not interfere with the result). So, if the reference plane is the foremost surface (surface 4) and five points are selected on that surface, first segment includes 3906 points, second segment 6542 points, third segment 1997 points, and fourth segment 9487 points (note that the source of 5 points is surface 4) [10]. Matlab coding flowchart for this application is given in **Figure 15**.

Matlab Code of Flowchart in **Figure 15** has been given in Appendix A.

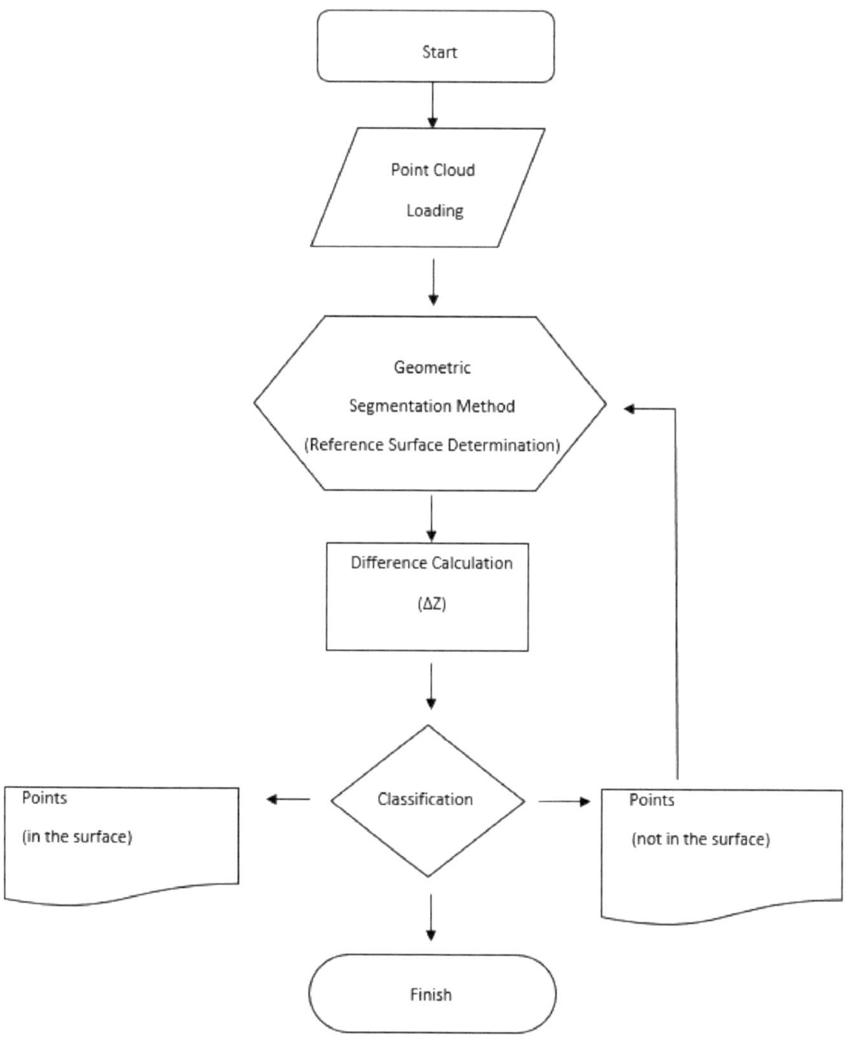

Figure 15.
Flowchart of Matlab coding for point segmentation.

3.2 Object segmentation

In this application, it is aimed to obtain the corner coordinate values of window gaps with geometric segmentation, which are not present in the mobile laser scanning data in order to use these values for automated visualization of the building facade survey. The sharp corners of the window are both natural points and target points for the vertical conversion. For this reason, laser point cloud data is used for the detection of these points by geometric segmentation. Then, the window corner points obtained by this segmentation are compared numerically with the values in the original point cloud data data in order to perform the analysis and obtain the results of the study.

The point cloud data used in the study are obtained from a measurement which was made in the Balgat district of Ankara with a vehicle in the inventory of TOPCON Company, equipped for mobile mapping, with a speed ranging from 10 km/h to 20 km/h. Ladybug-5 camera is used to produce photogrammetric data. The measuring equipment is shown in **Figure 16** [11].

The point cloud data of study study area which are collected mobil laser scanner has shown in the **Figure 17**.

In this application, the data, which were obtained from cyclone program, were saved into the Tamyatay.txt file. Then, in order to distinguish the building points from the ground points on which we performed the survey process, the points which were 35 cm high from the lowest point of the ground have been determined on our raw data and the classification process was started.

Ground criteria was determined by the following Eq. (3) [12]:

$$Z_{1} = Z_{min} + zk \quad (3)$$

Here, the ground height value zk (ground thickness), which is asked to be entered into the code, can be changed. In order to distinguish the points more easily in the study, 35 cm was taken as the most appropriate value. Before starting the classification process, the scanned point clouds were plotted in Matlab as below the **Figure 18**.

For performing the classification process, first, a value of 1 was assigned to the class column of all points. Then, Z value was assigned to points whose Z value was less than the ground criteria. The ground points matrix Matrix_Ground was created by selecting the values which were assigned with 0 and its graph was plotted. The road's center axis $Axis_y$, $Axis_x$ was determined using the points having maximum

Figure 16.
The vehicle of application which was used to obtain point cloud data.

Figure 17.
Mobil point cloud data of application Array.

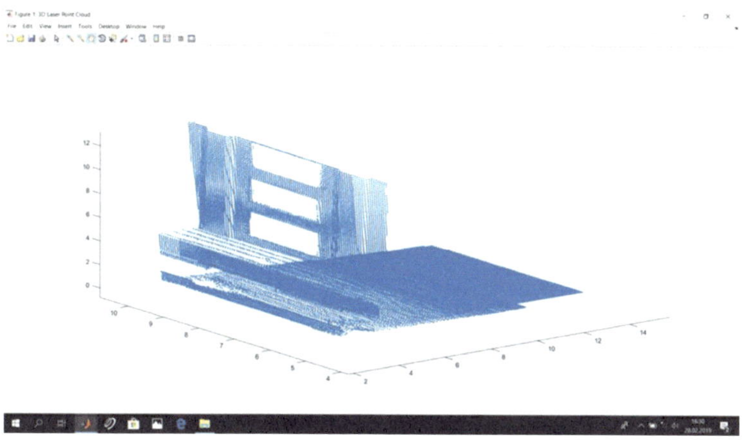

Figure 18.
Three-dimensional laser point cloud data in Matlab.

and minimum X and Y values from the ground points. When determining the center axis of the road, the axis was used on the points which were taken as $Z_{1=} Z$ and had a depth of Z_1 and above.

These processes were calculated with the following Eqs. (4) and (5):

$$Axis_y = Ground_{y_{min}} + \left(\frac{Ground_{y_{max}} - Ground_{y_{min}}}{2} \right) \quad (4)$$

$$Axis_x = Ground_{x_{min}} + \left(\frac{Ground_{x_{max}} - Ground_{x_{min}}}{2} \right) \quad (5)$$

Other than the values which were assigned to the class column with the value of 0 (ground point), the points assigned with class value 1 (building points) were saved as Matrix_Building_seg matrix and its graph was plotted in the **Figure 19**.

The building points were separated from the ground height of 35 cm, but a new classification was made on the basis of the above-mentioned road center axis to determine whether the building was on the left or right side of the road axis. For this purpose, a value of 1 was assigned to points smaller than $Axis_x$ (depth of the road axis) and a value of 2 was assigned to points greater than $Axis_x$. The points assigned with the value of 1 are shown in the Matrix_Building_Left matrix, and the points assigned with the value of 2 are shown in the Matrix_Building_Right matrix.

It is assumed that the building lays on the side with the high number of points. So, the number of points of the Matrix_Building_Left and Matrix_Building_Right matrices were calculated and the values of the Matrix_Building_Left matrix were written as Matrix_Building_Segment matrix where the building points were completely separated and then, the building points were plotted. Also, if there were no points exceeding the height limit (Z_1) on the right side of the road center axis, although this data was not included in our data, this classification step would have to be made in order to completely make a distinction of the building points in the **Figure 19** also.

After determining the building points, building points were transferred into a reference surface to draw the building facade survey. A mathematical filtering function was applied in this stage.

The segmentation works on the basis of surface-dependent height differences within the maximums and minimums of the surface function that forms the mathematical model. Surface equation is expressed as Eq. (6) [13]:

$$ax_n + by_n + cz_n + d = 0 \tag{6}$$

where n is the number of points.

An exponential filtering method is used as the filtering function. Filtering function is determined as Eq. (7):

$$\mathrm{f(x)} = e^{\Delta Building} = \frac{1}{e^{\Delta Depth}} \tag{7}$$

Where $\Delta X_{Building}$ and Δ_{Depth} were calculated with the following Eqs. (8) and (9):

$$\Delta X_{Building} = \frac{d + bZ_{Building} + aY_{Building}}{c} \tag{8}$$

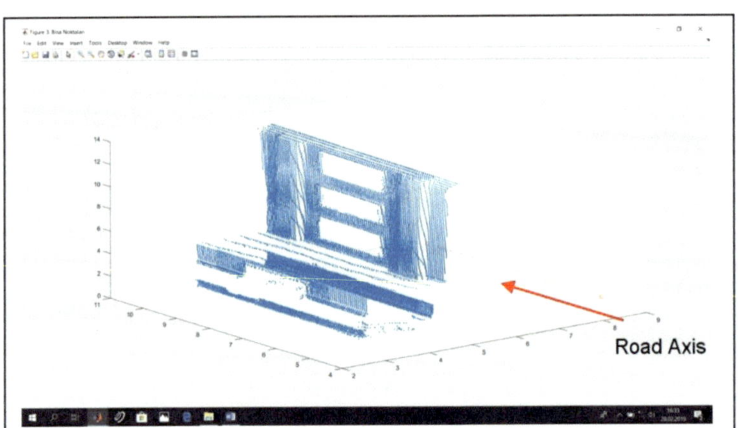

Figure 19.
Road center Axis and point cloud data without point cloud of road in Matlab.

$$\Delta_{Depth} = X_{Building} - X_{max} \tag{9}$$

a, b, c, d represent plane equation parameters, $\Delta X_{Building}$ is the depth of building points to the reference surface, Δ_{Depth} is depth differences of building points to the reference surface, $e^{\Delta Bina}$ is result values of the function, $e^{\Delta Depth}$ is exponential values of differences of depth, and $Eksen_y$ represents the y value of road center axis.

For the segmentation process, the building reference surface points were determined at first. Then, the limits of the reference surface were resolved by using the minimum and maximum points of Z (height) and Y (length) of the building points and a rectangular surface was created. Only the maximum value of X (depth) was used. This is due to the fact that the bottom window (Window 1) of the building, which we were to transfer to the surface is located more on the backside than the other three windows. Also, when the minimum value of X is taken, there is a possibility that the points behind the bottom window with the smaller depth values might interfere with the bottom window and building points. The specified reference surface points are written into a Building_Reference_Surface_Points.txt file in the **Table 2**. The Surface graphic has shown in the **Figure 20**.

After the reference surface points are determined, the reference plane parameters of the building are calculated using these points. The reference surface is drawing in Matlab has shown in the **Figure 21**, then the PRight matrix is created and plotted. It is written into a Building_Reference_Plane_Par.txt file.

X (m.)	Y (m.)	Z (m.)
5.424	10.537	13.598
5.424	4.762	13.598
5.424	10.537	1.014
5.424	4.762	1.014

Table 2.
Building reference surface points.

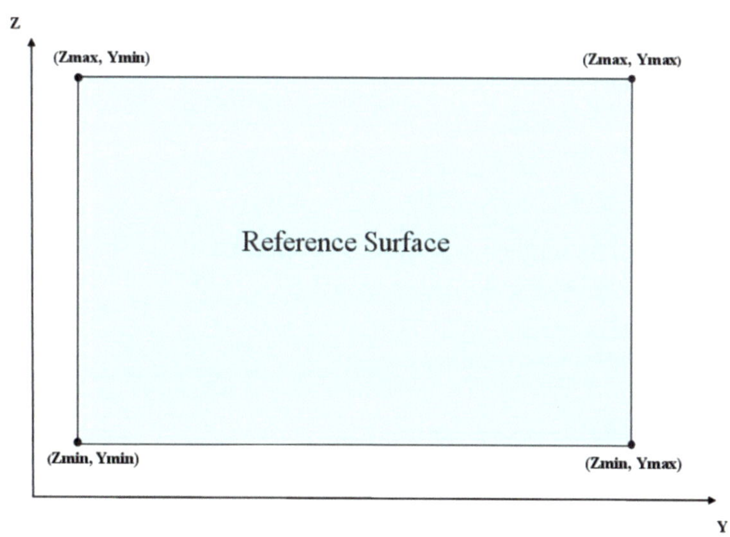

Figure 20.
Building reference surface points.

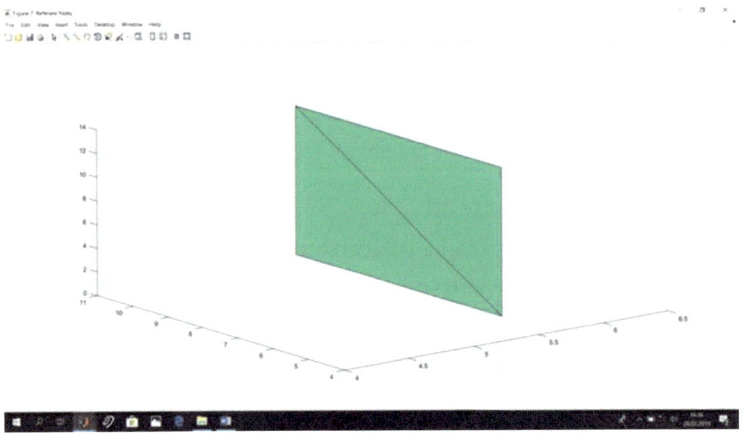

Figure 21.
Reference surface.

The following Eqs. (10–13) are used in the process of adjustment:

$$I = [4, 1] \tag{10}$$

$$N = BY^T BY \tag{11}$$

$$n = N^{-1} I \tag{12}$$

$$X = N^{-1}[I] \tag{13}$$

It represents unit matrix, BY is the matrix of detected maximum and minimum surface points, N is the normal equation coefficient matrix, n is the plain terms vector, and X represents the matrix of unknown values (surface parameters) in the **Table 3**.

Using the calculated parameters and the result values of the function ($e^{\Delta Building}$), Z and Y values of the building points are transferred to the reference plane. The X depth value ($Surface_{Depth}$) of the reference plane is calculated by averaging the maximum and minimum X values, and a $Surface_{Filter}$ matrix with the depth of the number of building points is created using the unit matrix. These processes are performed with the following Eqs. (14) and (15):

$$Surface_{Depth} = \frac{Surface_{X_{max}} + Surface_{X_{min}}}{2} \tag{14}$$

$$Surface_{Filter} = Surface_{Depth} \cdot I_{Unit} \tag{15}$$

To use the coordinate values obtained by mathematical filtering in the Geometric.m code, the new point cloud matrix was created and written to the

Plane Parameter Coefficients	
a	−0.0034287
b	−0.018451
c	0.039046
d	0.0071988

Table 3.
Reference plane parameters.

Geometric_Filter_Input.txt file for the references surface.The distance between points was determined as 25 cm, Geometric.m code was executed and 180 new points or openings were written and graphed by creating opening_output_txt file. As a result of this process, the window openings were filled vertically. This segmentation points have shown in the **Figure 22**.

Since the gaps were calculated in blocks with gap filtering in Matlab, a reclassification was performed to determine which window the dots belonged to. A function which is similar to the exponential filter function used in the transfer of building points to the reference plane was used. The windows which were drawn by using the coordinate values obtained from the result of point segmentation (distance between points is 14.5 cm) in red, surface segmentation in green, and conventional segmentation in magenta are shown in **Figures 23** and **24**.

To assign these points, the OP_Pen matrix with the number of lines (same as the building point numbers) and with a NaN value was created. Then, points within the 120 cm border were selected from the Matlab code matrix and printed on the OP_Pen matrix. In order to distinguish these assigned values from NaN values, the snip.m code was run once more. The points to be optimized for each window were differentiated and the files OP_PP1.txt, OP_PP2.txt, 0P_PP3.txt, OP_PP4.txt, which

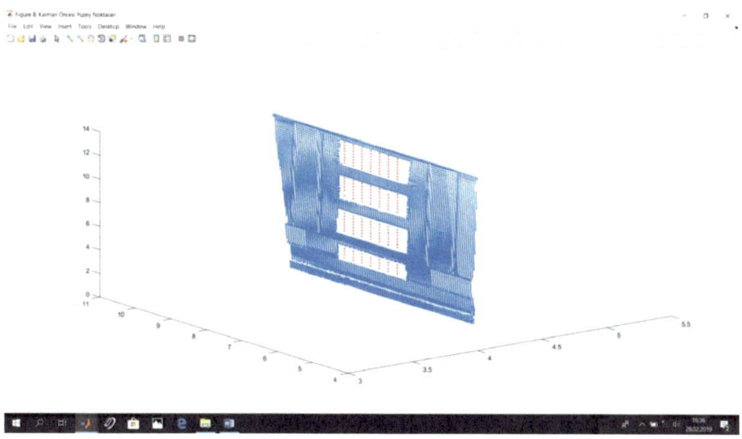

Figure 22.
Window points after geometric segmentation.

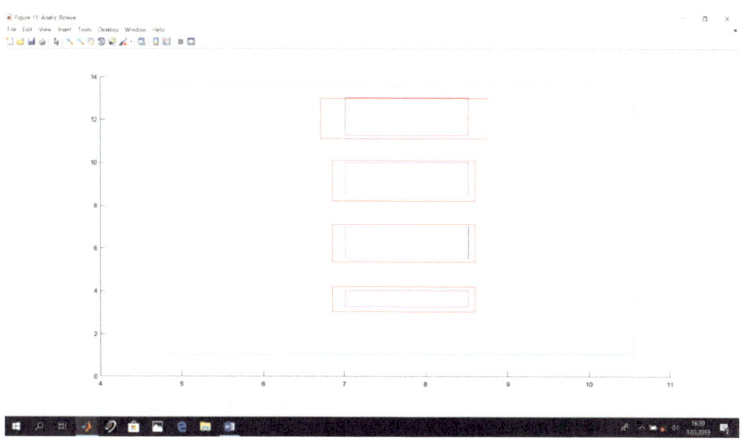

Figure 23.
Windows survey via two methods in MatLab.

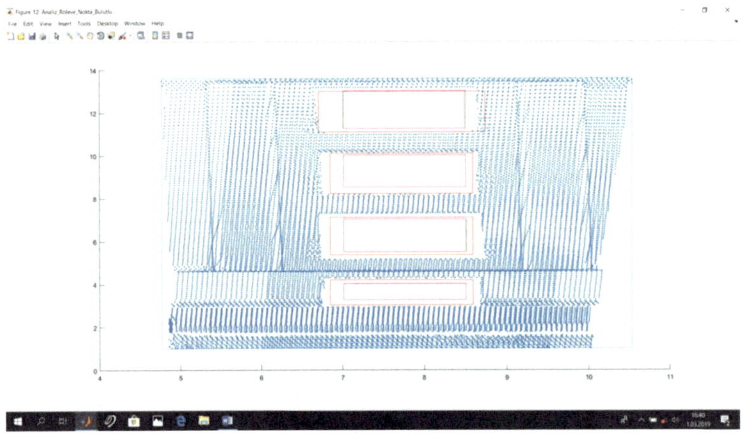

Figure 24.
Window gaps result of segmentation methods in survey point cloud in MatLab for window gaps.

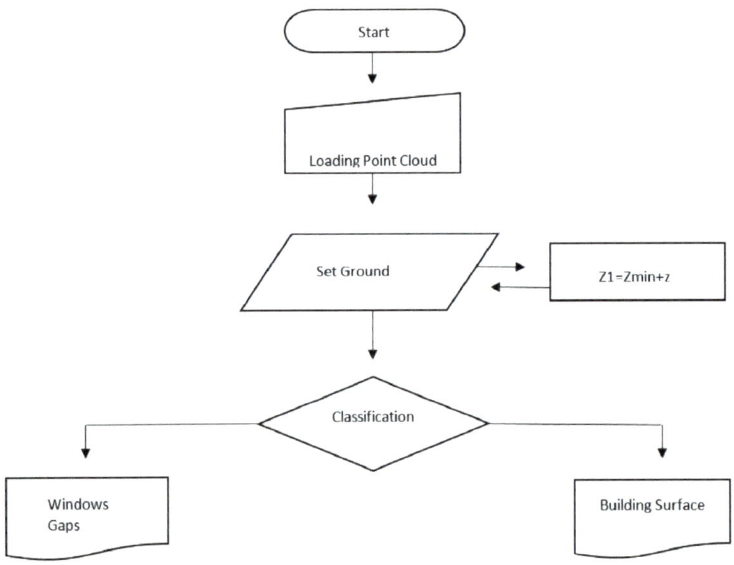

Figure 25.
Flowchart of Matlab coding for object segmentation.

had 2132, 646, 624, 383 points respectively, were created and the data points to be segmented were plotted in the **Figure 24**.

Matlab coding flowchart for this application is given in **Figure 25**.

Matlab Code of Flowchart in **Figure 25** has been given in Appendix B.

4. Conclusion

With this chapter, we observe that point cloud data need fundamental process flow as the fundamental steps of geometric segmentation by Matlab programming. Obviously, Matlab programming which depend on various algorithm by data structures. Geometric segmentation methods should be based on these alghorithm by programmers in the point cloud data. The process examples and some tips are given by Matlab coding in this chapter, which is made for some kind of point cloud segmentation flowcharts, and an approach to solve fundamental geometric segmentation is presented. Generaly, point cloud data coordinate system does not defined universal coordinate system. Thus, point cloud data listed disorderly by Cyclone 5.2 software are classified in a surface-based way, so it passes through a geometric segmentation before transformation to universal coordinate system.

In the potential, upcoming studies, radiometric and hiybrid segmentation methods might be used in Matlab coding. More accurate results might be obtained by using another method than geometric segmentation technique or by adding a third filter to geometric filtering or other segmentation methods whic are mentioned in this chapter.

Appendix A

```
clear all;
clc;
format long g;
load 'YUZEY1.txt';
load 'ham_data.txt';

I=ones(4,1);
N=transpose(YUZEY1)*(YUZEY1);
n=inv(N)*I;
X=inv(N)*(I);
[t,k]=size(ham_data);
int i;
int j;
i=1;
j=1;
matris_X=ones(t,1);
matris_Y=ones(t,1);
matris_Z=ones(t,1);

for j=1:k;
  i=1:t;
    c=ham_data(i,j);
    if j==1
      matris_X(i,1)=c;
    elseif j==2
        matris_Y(i,1)=c;
      elseif j==3
          matris_Z(i,1)=c;
        end
end
a=X(1,1);
b=X(2,1);
c=X(3,1);
d=X(4,1);
T=[ham_data];

[ns,ss]=size(T);

for i=1:ns;
```

```
   Matris_Oran(i,1) =(abs(a*(T(i,1)+ b*T(i,2)+ c*T(i,3)+ d)))/sqrt(a*a+b*b+c*c);
end

Kesme_Kriteri = std(Matris_Oran);
%------------------------------------------------------------------------------------
%------------------------------------------------------------------------------------
%------------------------------------------------------------------------------------

kes = Kesme_Kriteri;
gebze = max(Matris_Oran);
cayir = min(Matris_Oran);
yuksek = mean(Matris_Oran);
%sayi = ((Ayrac-1)/kes);
for i=1:ns;
   Matris_YOran(i,1) =(Matris_Oran(i,1)/yuksek);
end

ayar1 = max(Matris_YOran);
ayar2 = min(Matris_YOran);
ayar3 = mean(Matris_YOran);
ayar4 = std(Matris_YOran);

for i=1:ns;
   Matris_KOran(i,1) =exp(Matris_YOran(i,1));
end

dayar1 = max(Matris_KOran);
dayar2 = min(Matris_KOran);
dayar3 = mean(Matris_KOran);
dayar4 = std(Matris_KOran);

%yüzey sayisinin belirlenmesi------------------------------------------------------

say = dayar1 - dayar2;
say = round (say);
yuzey_sayisi= say;

%----kesme kriteri ve sýnýflandýrma---------------------------------------

kes=dayar2 + 0.25;
for i=1:ns;

    if (Matris_KOran(i,1) < kes)
                Matris_KEGIT(i,1) = 1;
    else
                Matris_KEGIT(i,1) = 0;
   end

end

for i=1:ns;
   if ((2.25 < Matris_KOran(i,1))&&( Matris_KOran(i,1) < 3.25))
            Matris_KEGIT(i,1) = 2;
```

```
    end
  end

  for i=1:ns;
    if ((3.25 <= Matris_KOran(i,1))&&( Matris_KOran(i,1) < 4.25))
                Matris_KEGIT(i,1) = 3;

    end
  end

  for i=1:ns;
    if ((4.25 <= Matris_KOran(i,1))&&( Matris_KOran(i,1) < 5.50))
                Matris_KEGIT(i,1) = 4;
    end
  end

%-----------------------------------------------------------------------------
%-----------------------------------------------------------------------------

KY=ham_data(:,2);
KX=ham_data(:,1);
KZ=ham_data(:,3);
%Fig1=plot(KX,KY,KZ,'–rs');

T=[T,Matris_Oran,Matris_YOran,Matris_KOran,Matris_KEGIT];

Fig2 = plot(Matris_KOran,'-.or');
```

Appendix B

```
 clc;
clear;
format long g;

%Point cloud data loading

[dosyaadi,dosyayolu] = uigetfile(...
   {'*.dat;*.txt;*.xyz;*.pts','Lazer Veri Dosyaları...(*.dat,*.txt,*.xyz,*.pts)';
   '*.dat', 'Data_Dosyalar (*.dat)';...
   '*.txt', 'Txt_Dosyalar (*.txt)';...
   '*.xyz', 'Nokta_Dosyalar (*.xyz)';...
   '*.pts', 'Nokta Bulutu_Dosyalar (*.pts)';
   '*.*', 'Tüm Dosyalar (*.*)'},...
   'Bir Lazer Tarama Veri Dosyası Seçiniz:');

 if dosyaadi~=0
h=waitbar(0,'Lazer Verisi Yükleniyor');
for i=1:10

 Ham_veri_Matris = load([dosyayolu,dosyaadi]);

 [ns, ss] = size(Ham_veri_Matris);
```

```
  waitbar(i/10,h);
end

  close(h);
end

  %Segmentation starting

  X =Ham_veri_Matris(:,1);
Y =Ham_veri_Matris(:,2);
Z =Ham_veri_Matris(:,3);

  figure('Name','3D Laser Point Cloud','NumberTitle','on')
scatter3(X,Y,Z,'.');

  str1=num2str(ns);
uiwait(msgbox({'Toplam Tarama Nokta Sayısı',[str1]},'Success'));
fprintf('Toplam Tarama Nokta Sayısı %d\n',ns);

  matris_sinif=ones(ns,1);
Matris_segmentation = [X,Y,Z,matris_sinif];

  %Elevetion extraction
zemin=min(Z);

  %Elevation determined
  ifade={'Zemin Yüksekliği Değerini Giriniz !'};
baslik='Zemin Kalınlığı (birim m)';
normal={'0.35'};
zemin_kln=inputdlg(ifade,baslik,1,normal);
zemin_kln=str2double(zemin_kln);
Z1= zemin+zemin_kln;

  k=0;
for i=1:ns

      if (Ham_veri_Matris(i,3) <=Z1)
            Matris_segmentation(i,4)=0;
      k=k+1;
     end
end

for i=1:ns

    if (Matris_segmentation(i,4)== 0)

        Matris_Yer(i,1)=Ham_veri_Matris(i,1);
      Matris_Yer(i,2)=Ham_veri_Matris(i,2);
      Matris_Yer(i,3)=Ham_veri_Matris(i,3);

    end

end
```

```
 Yer_X=Matris_Yer(:,1);
Yer_Y=Matris_Yer(:,2);
Yer_Z=Matris_Yer(:,3);

 Yer_X = nonzeros(Yer_X);
Yer_Y = nonzeros(Yer_Y);
Yer_Z = nonzeros(Yer_Z);

 Matris_Yer=[Yer_X Yer_Y Yer_Z];

 [yns,yss]=size(Matris_Yer);

 figure('Name','Zemin Noktaları','NumberTitle','on')
scatter3(Yer_X,Yer_Y,Yer_Z,'.');
hold on;

 str2=num2str(yns);
msgbox({'Zemin Noktası Sayısı',[str2]},'Success');
str3=num2str(ns-yns);
msgbox({'Bina Noktası Sayısı',[str3]},'Success');

 fprintf('Zemin Noktası Sayısı %d\n',yns);
fprintf('Bina Noktası Sayısı %d\n',ns-yns);
 %Base axes determination

 Yer_y_min=min(Yer_Y);
Yer_x_min=min(Yer_X);
Yer_y_max=max(Yer_Y);
Yer_x_max=max(Yer_X);

 Eksen_y=Yer_y_min+((Yer_y_max-Yer_y_min)/2);
Eksen_x=Yer_x_min+((Yer_x_max-Yer_x_min)/2);
Line1=[Eksen_x Yer_y_min];
Line2=[Eksen_x Yer_y_max];
Point_X=[Eksen_x
     Eksen_x];

 Point_Y=[Yer_y_min
     Yer_y_max];
Point_Z=[Z1
     Z1];

 %Segmentation matrix refreshing

 Matris_seg=zeros(size(Matris_segmentation));
k=0;
t=0;
for i=1:ns
 if (Matris_segmentation(i,4)==1)

     Matris_seg(i,1)= Matris_segmentation(i,1);
    Matris_seg(i,2)= Matris_segmentation(i,2);
    Matris_seg(i,3)= Matris_segmentation(i,3);
```

```
      Matris_seg(i,4)= Matris_segmentation(i,4);
      k=k+1;
     else if (Matris_segmentation(i,4)==0)
      Matris_seg(i,1)= 0;
      Matris_seg(i,2)= 0;
      Matris_seg(i,3)= 0;
      Matris_seg(i,4)= 0;
      t=t+1;
     end

   end
 end

  %Building Points determinated

  Bina_X=nonzeros (Matris_seg(:,1));
 Bina_Y=nonzeros (Matris_seg(:,2));
 Bina_Z=nonzeros (Matris_seg(:,3));

  Matris_Bina=[Bina_X,Bina_Y,Bina_Z];

  [bns,bss]=size(Matris_Bina);

  sinif=ones(bns,1);

  Matris_Bina_seg=[Matris_Bina,sinif];
  %Building Points drawing

  figure('Name','Bina Noktaları','NumberTitle','on')

  scatter3(Bina_X,Bina_Y,Bina_Z,'.');hold on
 plot3(Point_X,Point_Y,Point_Z, '-.o');
 hold on
 grid on

  %Building Points segmented

  sol=0;
 sag=0;
 for i=1:bns

      if (Matris_Bina(i,1) < Eksen_x )
         Matris_Bina_seg(i,4)= 1;
         sol=sol+1;
     else if (Matris_Bina(i,1) > Eksen_x )
         Matris_Bina_seg(i,4)= 2;
         sag=sag+1;
        end
    end

 end

  Matris_Bina_sol=zeros(bns,3);
```

```
Matris_Bina_sag=zeros(bns,3);

  for i=1:bns
          if (Matris_Bina_seg(i,4)==2)
          Matris_Bina_sag(i,1)= Matris_Bina_seg(i,1);
          Matris_Bina_sag(i,2)= Matris_Bina_seg(i,2);
          Matris_Bina_sag(i,3)= Matris_Bina_seg(i,3);
          else
          Matris_Bina_sag(i,1)= 0;
          Matris_Bina_sag(i,2)= 0;
          Matris_Bina_sag(i,3)= 0;
          end
end

  Matris_Bina_Sag_X=Matris_Bina_sag(:,1);
Matris_Bina_Sag_Y=Matris_Bina_sag(:,2);
Matris_Bina_Sag_Z=Matris_Bina_sag(:,3);

  Matris_Bina_Sag_X=nonzeros(Matris_Bina_Sag_X);
Matris_Bina_Sag_Y=nonzeros(Matris_Bina_Sag_Y);
Matris_Bina_Sag_Z=nonzeros(Matris_Bina_Sag_Z);
Matris_Bina_Sag=[Matris_Bina_Sag_X,Matris_Bina_Sag_Y,Matris_Bina_Sag_Z];

for i=1:bns
         if (Matris_Bina_seg(i,4)==1)
         Matris_Bina_sol(i,1)= Matris_Bina_seg(i,1);
         Matris_Bina_sol(i,2)= Matris_Bina_seg(i,2);
         Matris_Bina_sol(i,3)= Matris_Bina_seg(i,3);
         else
         Matris_Bina_sol(i,1)= 0;
         Matris_Bina_sol(i,2)= 0;
         Matris_Bina_sol(i,3)= 0;
         end
end

  Matris_Bina_Sol_X=Matris_Bina_sol(:,1);
Matris_Bina_Sol_Y=Matris_Bina_sol(:,2);
Matris_Bina_Sol_Z=Matris_Bina_sol(:,3);

  Matris_Bina_Sol_X=nonzeros(Matris_Bina_Sol_X);
Matris_Bina_Sol_Y=nonzeros(Matris_Bina_Sol_Y);
Matris_Bina_Sol_Z=nonzeros(Matris_Bina_Sol_Z);

  Matris_Bina_Sol=[Matris_Bina_Sol_X,Matris_Bina_Sol_Y,Matris_Bina_Sol_Z];

  %Left side – right side deciding

  [bnsol,bnssol] = size(Matris_Bina_Sol);
[bnsag,bnssag] = size(Matris_Bina_Sag);

  if bnsol > bnsag
    msgbox('Bina, yol eksenine göre Sol taraftadır !','Success');
    [bina,sut_bina]=size(Matris_Bina_Sol);
```

```
    Matris_sinif=ones(bina,1);
    Matris_Bina_Segment=[Matris_Bina_Sol,Matris_sinif];

  %Segmented Building Points drawing
figure('Name','Bina Noktaları','NumberTitle','on')
scatter3(Matris_Bina_Sol_X,Matris_Bina_Sol_Y,Matris_Bina_Sol_Z,'.');
hold on;

else
    msgbox('Bina, yol eksenine göre Sağ taraftadır !','Success');
    [bina,sut_bina]=size(Matris_Bina_Sag);
    Matris_sinif=2*ones(bina,1);
    Matris_Bina_Segment=[Matris_Bina_Sag,Matris_sinif];

end

  %Referance surface determination

  Bina_yuzey_X_max = max(Matris_Bina_Segment(:,1));
Bina_yuzey_X_min = min(Matris_Bina_Segment(:,1));

  Bina_yuzey_Z_max = max(Matris_Bina_Segment(:,3));
Bina_yuzey_Z_min = min(Matris_Bina_Segment(:,3));

  Bina_yuzey_Y_max = max(Matris_Bina_Segment(:,2));
Bina_yuzey_Y_min = min(Matris_Bina_Segment(:,2));

  BY1=[Bina_yuzey_Z_max,Bina_yuzey_Y_max,Bina_yuzey_X_max,1];
BY2=[Bina_yuzey_Z_max,Bina_yuzey_Y_min,Bina_yuzey_X_max,1];
BY3=[Bina_yuzey_Z_min,Bina_yuzey_Y_max,Bina_yuzey_X_max,1];
BY4=[Bina_yuzey_Z_min,Bina_yuzey_Y_min,Bina_yuzey_X_max,1];

  BG1=[Bina_yuzey_Z_max,Bina_yuzey_Y_max,Bina_yuzey_X_max];
BG2=[Bina_yuzey_Z_max,Bina_yuzey_Y_min,Bina_yuzey_X_max];
BG3=[Bina_yuzey_Z_min,Bina_yuzey_Y_max,Bina_yuzey_X_max];
BG4=[Bina_yuzey_Z_min,Bina_yuzey_Y_min,Bina_yuzey_X_max];

  BG=[BG1
    BG2
    BG3
    BG4];

  BGGX=BG(:,1);
BGGY=BG(:,2);
BGGZ=BG(:,3);

BY=[BY1
    BY2
    BY3
    BY4];

 %Reference surface Points
```

```
GZ=[Bina_yuzey_Z_max
   Bina_yuzey_Z_max
   Bina_yuzey_Z_min
   Bina_yuzey_Z_min];

GY=[Bina_yuzey_Y_max
   Bina_yuzey_Y_min
   Bina_yuzey_Y_max
   Bina_yuzey_Y_min];

GX=[Bina_yuzey_X_max
   Bina_yuzey_X_max
   Bina_yuzey_X_max
   Bina_yuzey_X_max];

%Reference surface drawing

 figure('Name','Bina En Büyük Referans Yüzey Noktaları','NumberTitle','on')
scatter3(GX,GY,GZ,'.');
hold on;

 figure('Name','Bina En Büyük Referans Yüzeyi','NumberTitle','on')
plot3(GX,GY,GZ);
grid on;

GZ=[Bina_yuzey_Z_min
   Bina_yuzey_Z_max
   Bina_yuzey_Z_max];

KZ=[Bina_yuzey_Z_max
   Bina_yuzey_Z_min
   Bina_yuzey_Z_min];

GY=[Bina_yuzey_Y_min
   Bina_yuzey_Y_max
   Bina_yuzey_Y_min];

KY=[Bina_yuzey_Y_max
   Bina_yuzey_Y_min
   Bina_yuzey_Y_max];

GX=[Bina_yuzey_X_max
  Bina_yuzey_X_max
   Bina_yuzey_X_max];

 figure('Name','Referans Yüzey','NumberTitle','on')
fill3(GX,GY,GZ,GX);
grid on
hold on
fill3(GX,KY,KZ,GX)
grid on
hold on
```

```
 dlmwrite ('Bina_Referans_Yuzey_noktaları.txt',BY,'delimiter','\t');
I=ones(4,1);
N=transpose(BY)*BY;
n=pinv(N)*I;
X=pinv(N)*(I);

 %Reference surface parameters

 a=X(1,1);
b=X(2,1);
c=X(3,1);
d=X(4,1);

PSag=[a
  b
  c
  d];

 dlmwrite('Bina_Referans_Düzlem_Par.txt',PSag,'delimiter','\t');
%Building surface segmentation

for i=1:bina
  Derinlik_Bina(i,1)=(-(d+b*Matris_Bina_Segment(i,2)+a*Matris_Bi-
  na_Segment(i,3))/c);
  Delta_Derinlik(i,1)= Derinlik_Bina(i,1)- Bina_yuzey_X_max;
  DDEXP_Bina(i,1) = 1/(exp(Delta_Derinlik(i,1)));
  Egim_Acisi_Bina(i,1) = atan((Matris_Bina_Segment(i,2)-Z1)/(Matris_
Bina_Segment(i,1)- Eksen_y));
end

 Yuzey_Der=(Bina_yuzey_X_max+Bina_yuzey_X_min)/2;
Filtre_Yuzey = Yuzey_Der*ones(bina,1);
Matris_Bina_Segment=[Matris_Bina_Segment,Derinlik_Bina,Delta_Derinlik,
DDEXP_Bina,Filtre_Yuzey];

 KOD=[Matris_Bina_Segment(:,2) Matris_Bina_Segment(:,3) Matris_
Bina_Segment(:,8)];
dlmwrite('Geo_Filitre_Giriş.txt',KOD,'delimiter','\t');

 %Geometric Segmentation start
run Geometric.m;
%Geometric Segmentation finish

KSM = dlmread('bosluk_output.txt');

hold on;

 figure('Name','Geometric Öncesi Yüzey Noktaları','NumberTitle','on')
scatter3(Matris_Bina_Segment(:,8),Matris_Bina_Segment(:,2),Matris_
Bina_Segment(:,3),'.');hold on
scatter3(KSM(:,3),KSM(:,1),KSM(:,2),'.','r');
hold on;
```

```
 Ref_Nok_X=Point_X/2;
Ref_Nok_Y=Point_Y/2;
Ref_Nok_Z=Z1;

 %Building flat number info ask

 ifade1={'Bina Kaç Katlı ?'};
normal1={'4'};
baslik1='Bina Kat Sayısı';
satir1=1;
cevap1=inputdlg(ifade1,baslik1,satir1,normal1);
Bina_Kat_Sayisi=str2double(char(cevap1(1,1)));

 %Flat Classing

 [kfns,kfss]=size(KSM);
Egim_Acisi_Geo=ones(kfns,1);
EXP_Kal=ones(kfns,1);
for i=1:kfns
   Egim_Acisi_Geo(i,1) = atan(KSM(i,2)/10);
   EXP_Kal(i,1)=exp(Bina_Kat_Sayisi*Egim_Acisi_Geo(i,1));
end

 std_Kal=std(EXP_Kal);
max_Kal=max(EXP_Kal);
min_Kal=min(EXP_Kal);
KAL_Seg_Deg = round(std_Kal,0)/(Bina_Kat_Sayisi*0.5);
Seg_Kal=[KSM Egim_Acisi_Geo EXP_Kal];

 for i=1:Bina_Kat_Sayisi
   eval(sprintf('Pen%d = NaN(kfns,3)', i));
end

for j=Bina_Kat_Sayisi:-1:1
   for i=1:kfns
      if Seg_Kal(i,5)>KAL_Seg_Deg*2^(j-2) && Seg_Kal(i,5)<KAL_Seg_Deg*2^
      (j-1)
      eval(sprintf('Pen%d(i,1)=Seg_Kal(i,1)', j));
      eval(sprintf('Pen%d(i,2)=Seg_Kal(i,2)', j));
      eval(sprintf('Pen%d(i,3)=Seg_Kal(i,3)', j));
     end

   end
PP=snip(eval(sprintf('Pen%d', j)),nan);
eval(sprintf('PP%d = PP', j));

  PX=[max(eval(sprintf('PP%d(:,1)', j)))
   min(eval(sprintf('PP%d(:,1)', j)))
   min(eval(sprintf('PP%d(:,1)', j)))
   max(eval(sprintf('PP%d(:,1)', j)))
   max(eval(sprintf('PP%d(:,1)', j)))];

     eval(sprintf('aax%d=PX(1,1)',j));
```

```
eval(sprintf('bbx%d=PX(2,1)',j));

eval(sprintf('PP%dX = PX', j));

PY=[min(eval(sprintf('PP%d(:,2)', j)))
  min(eval(sprintf('PP%d(:,2)', j)))
  max(eval(sprintf('PP%d(:,2)', j)))
  max(eval(sprintf('PP%d(:,2)', j)))
  min(eval(sprintf('PP%d(:,2)', j)))];

  eval(sprintf('aay%d = PY(3,1)',j));
 eval(sprintf('bby%d = PY(1,1)',j));

  eval(sprintf('PP%dY = PY', j));

PZ=[eval(sprintf('PP%d(1,3)', j))
  eval(sprintf('PP%d(1,3)', j))
  eval(sprintf('PP%d(1,3)', j))
  eval(sprintf('PP%d(1,3)', j))
  eval(sprintf('PP%d(1,3)', j))];
  eval(sprintf('PP%dZ = PZ', j));

end

 KODX=[max(KOD(:,1))
  min(KOD(:,1))
  min(KOD(:,1))
  max(KOD(:,1))
  max(KOD(:,1))];

 KODY=[min(KOD(:,2))
  min(KOD(:,2))
  max(KOD(:,2))
  max(KOD(:,2))
  min(KOD(:,2))];

 KODZ=[KOD(1,3)
  KOD(1,3)
  KOD(1,3)
  KOD(1,3)
  KOD(1,3)];

 %Results drawing

 figure('Name','Röleve','NumberTitle','on')
for i=1:Bina_Kat_Sayisi
line(eval(sprintf('PP%dX', i)),eval(sprintf('PP%dY', i)),eval(sprintf('PP%dZ', i)),'Color','red');hold on
end
line(KODX,KODY,KODZ);hold on
```

Author details

Bahadır Ergün* and Cumhur Şahin
Department of Geomaics Engineering, Gebze Technical University, PK 141, Gebze 41400 Kocaeli, Türkiye

*Address all correspondence to: bergun@gtu.edu.tr

References

[1] Lichti D, Gordon SJ. Error Propagation in directly georeferenced terrestrial laser scanner point clouds for cultural heritage recording. In: Proceeding of FIG Working Week; 22-27 May 2004; Athens, Greece; 2004

[2] Ergun B. A novel 3D geometric object filtering function for application in indoor area with terrestrial laser scanning data. Optics & Laser Technology. 2010;**42**: 799-804

[3] Lerma JL, Biosca JM. Segmentation and filtering of laser scanner data for cultural heritage. In: Proceeding of CIPA 2005 XX International Symposium; 26 September–01 October, 2005; Torino, Italy; 2005

[4] Dold C, Brenner C. Automatic matching of terrestrial scan data as a basis for the generation of detailed 3D city models. In: Proceeding of International Archives of Photogrammetry, Remote Sensing and Spatial Information Sciences; 12-23 July 2004; Istanbul, Turkey; 35(B3) pp. 1091-1096

[5] Pu S, Vosselman S. Automatic extraction of building features from terrestrial laser scanning. In: Proceeding of International Archives of the Photogrammetry, Remote Sensing and Spatial Information Sciences; 25-27 September 2006; Dresden, Germany; 36(5):5 pages

[6] Biosca JM, Lerma JL. Unsupervised robust planar segmentation of terrestrial laser scanner point clouds based on fuzzy clustering methods. In: Proceeding of ISPRS Journal of Photogrammetry and Remote Sensing, 3-11 July 2008; Beijing, China; **63**(1):84-98

[7] Bauer J, Karner K, Schindler K, Klaus A, Zach C. Segmentation of building models from dense 3D point-clouds. In: Proceedings of 27th Workshop of the Austrian Association for Pattern Recognition; 5-6 June 2003; Laxenburg, Austria. 2003; pp. 253-258

[8] Boulaassal H, Landes T, Grussenmeyer P, Tarsha-Kurdi F. Automatic segmentation of building facades using terrestrial laser data. In: Proceedings of International Archives of the Photogrammetry, Remote Sensing and Spatial Information Sciences, Workshop on Laser Scanning 2007 and SilviLaser 2007; Espoo, Finland. Vol. 36 (3/W52). 2007. pp. 65-70

[9] Boulaassal H, Landes T, Grussenmeyer P. Automatic extraction of planar clusters and their contours on building façades recorded by terrestrial laser scanner. International Journal of Architectural Computing. 2009;**7**(1): 1-20

[10] Sahin C. Planar segmentation of indoor terrestrial laser scanning point clouds via distance function from a point to a plane. Optics and Lasers in Engineering. 2015;**64**:23-31. DOI: 10.1016/j.optlaseng.2014.07.007

[11] Ergun B, Sahin C, Aydin A. Two-dimensional (2-D) Kalman segmentation approach in mobile laser scanning (MLS) data for panoramic image registration. Lasers in Engineering. 2020;**48**:121-150

[12] Grewal MS, Andrews A. Kalman Filtering Theory and Practice Using MATLAB. 2nd ed. Canada: Wiley; 2001

[13] Ay E. 3D geometric object fıltering application in terrestrial laser scanning data. [Master's thesis]. Kocaeli, Türkiye: Gebze Technical University; 2009

Section 4

Bluetooth Applications

Chapter 4

Bluetooth Low Energy Applications in MATLAB

Septimiu Mischie

Abstract

This chapter presents Bluetooth Low Energy (BLE) applications in MATLAB. Through these applications we acquire measurement data from BLE compatible sensors to PC. The sensors are CC2541 Keyfob and CC2650 Sensor Tag. The first one contains an accelerometer and a temperature sensor while the second one contains more sensors, but inertial sensors and magnetometer are invoked. The PC should be equipped with a general USB BLE adapter. The most important steps for implementing a BLE application are presented: scanning, connecting, configuring and data reading. Following this, more detailed applications are presented: a wireless sensor network for temperature measurement with three Keyfob-based nodes, an application that displays in real time accelerometer data and a heading computed method using either the gyroscope or the magnetometer of CC2650 Sensor Tag. The most important MATLAB elements that are used to implement these applications are different types of variables such as structure, table and object, methods to implement endless loops and real-time display of acquired data and using quaternions to handle 3D orientation of a device.

Keywords: MATLAB, Bluetooth low energy, temperature sensors, movement sensors, callback function, quaternion, 3D orientation

1. Introduction

MATLAB represents a programming language that is used for designing, simulating and testing of different technical systems [1]. This chapter provides examples of Bluetooth low energy (BLE) applications implemented in MATLAB. In this section, the main aspects regarding developing a BLE MATLAB application are presented. First of all, basics about BLE technology are presented [2–5]. BLE means exchange data between two or more devices by radio waves over short distances. Mainly, a BLE device can be scanner or advertiser. The advertiser signals its presence by sending its name and address. The scanner finds advertiser devices and can connect to one or more of them. Then the advertiser becomes a server and can send data to the scanner which is now a client. According to BLE architecture, the server can offer services to the client. Some examples of services are battery service, accelerometer service and heart rate measurements. Each service contains more characteristics. The most important attribute of a characteristic is its value, which in general represents sensor data. In addition, a characteristic has one or more of the following properties: read, write and notify.

Starting with 2019b release, MATLAB has introduced a set of functions that allow a simple implementation of BLE application [6]. The minimum setup involves a laptop having an embedded BLE adaptor or a desktop having an USB BLE adapter and some BLE compatible devices.

Scanning for BLE devices can be done using *blelist* function. Connecting to one device can then be done by *ble*() function, as in the examples of **Figure 1**.

It can be seen that three BLE devices have been discovered. The connection with Keyfobde99 is achieved, and b1 is a *ble* object having 5 fields, the last two being Services and Characteristics so in previous paragraph was presented. Then, in order to get data from a BLE device, the characteristics have to be accessed. **Figure 2** presents the last part of the Characteristics variable.

A characteristic can be accessed either by service name and characteristic name or by service universal unique identifier (UUID) and characteristic (UUID) [6]. According to the information from **Figure 2** it can be seen that only the second option can be used because there are more Custom services or characteristics.

```
>> list=blelist('timeout',15)

list =

  3×5 table

    Index            Name                 Address          RSSI    Advertisement
    _____    __________________    ______________    ____    _____________

      1      "CC2650 SensorTag"    "B0B448BD7E05"    -50      [1×1 struct]
      2      "Multi-Sensor"        "1804EDBEF596"    -56      [1×1 struct]
      3      "Keyfobde99"          "84DD20C50B29"    -71      [1×1 struct]

>> b1=ble("84DD20C50B29")

b1 =

  ble with properties:

               Name: "Keyfobde99"
            Address: "84DD20C50B29"
          Connected: 1
           Services: [10×2 table]
    Characteristics: [29×5 table]

Show services and characteristics

>>
```

Figure 1.
Using blelist *and* ble() *functions.*

```
>> table=b1.Characteristics(18:26,1:4)

table =

  9×4 table

      ServiceName         ServiceUUID    CharacteristicName    CharacteristicUUID
    _________________    ___________    __________________    __________________

    "Tx Power"             "1804"        "Tx Power Level"            "2A07"
    "Battery Service"      "180F"        "Battery Level"             "2A19"
    "Custom"               "FFA0"        "Custom"                    "FFA1"
    "Custom"               "FFA0"        "Custom"                    "FFA2"
    "Custom"               "FFA0"        "Custom"                    "FFA3"
    "Custom"               "FFA0"        "Custom"                    "FFA4"
    "Custom"               "FFA0"        "Custom"                    "FFA5"
    "Custom"               "FFA0"        "Custom"                    "FFA6"
    "Custom"               "FFA0"        "Custom"                    "FFA7"

>>
```

Figure 2.
A part of services and characteristics of a ble *variable.*

Furthermore, there is no information about the functionalities of these services or characteristics. Therefore, to start the application, some information is necessary that can be obtained using another application such as BLE Device Monitor [7] or just Wikipedia [8]. Thus, **Table 1** presents service name and characteristic name among to UUID for a part of the positions of **Figure 2**.

In order to access a characteristic, the *characteristic* () function can be used, as in **Figure 3**. Then, after defining x variable, the returned value can be read by function *read*(x).

Another powerful features is *DataAvailbleFcn* than can be assigned to a characteristics that has the Notify attribute. This can be done as in **Figure 4**. When a new data is available, this callback function is called.

Service Name	UUID	Characteristic Name	UUID
Accelerom. service	FFA0	Accelerometer enable	FFA1
		Accelerometer range	FFA2
		Accelerometer X coordinate	FFA3
		Accelerometer Y coordinate	FFA4
		Accelerometer Z coordinate	FFA5
		Period of Reading accelerometer data	FFA6
		Temperature	FFA7

Table 1.
UUID for a service and its characteristics.

```
>> enable=characteristic(b1,"FFA0","FFA1");
>> write(enable,1); %enable accelerometer
>> x=characteristic(b1,"FFA0","FFA3"); %x axis
>> data_x=read(x)

data_x =

    66

>>
```

Figure 3.
Using the characteristic() *function.*

```
Command Window
>> x.DataAvailableFcn=@displayCharacteristicData_x

x =

  Characteristic with properties:

             Name: "Custom"
             UUID: "FFA3"
       Attributes: "Notify"
      Descriptors: [2x3 table]
 DataAvailableFcn: displayCharacteristicData_x

Show descriptors

fx >>
```

Figure 4.
Create the callback function.

Each of the following sections contains an introduction where the basic function of the program is presented, which is then followed by the program and the results, mainly in graphical form.

2. BLE network sensors for temperature monitoring

This section presents a MATLAB application that uses three temperature sensors. CC2541 Keyfob [9] is a BLE compatible device that contains an accelerometer. Among its basic function, the accelerometer contains an 8-bit temperature sensor. To access the temperature sensor the accelerometer must be enabled first and then the temperature characteristic can be accessed according to **Table 1**. The period of reading temperature is 3 sec. according to the author publication [10].

At the beginning of the program a general scanning is executed and if none of the desired sensors are discovered the application is stopped through a suitable message on the screen. To do this, accessing the elements of a variable of table type, *list*, is performed.

Then, depending on the discovered number of sensors, which can be from one to any number (three in this application) the application gets temperature from them and displays it on a graphic. For this purpose, two structures, *s_enable* and *s_x*, having a variable number of fields have been created. Number of fields will be equal with the discovered number of sensors. The structure *s_x* is used to assign a callback function for each discovered sensor, too. Furthermore, the number of matrix of small axes which are generated by subplot function is equal with the number of discovered sensors.

The structure of the program is presented below.

```
%%%%%Measure the temperature by one, two or three
%%%%%Key-fob devices%%%%%%%%%%%%%%%
clear; close all;
global V_temp1;global V_temp2;global V_temp3
global V_time1;global V_time2;global V_time3
global N % the last N temperatures of each sensor
N=20;
V_temp1=zeros(1,N);V_temp2=zeros(1,N);V_temp3=zeros(1,N);
d=datetime;
V_time1=repmat(d,N,1);V_time2=repmat(d,N,1);V_time3=repmat(d,N,1);
list=blelist; %scan
L=size(list);
Nr=L(1);
disp(['Total number of BLE devices: ' num2str(Nr)])
if Nr==0
return
end
j=1;
for i=1:Nr
if (list.Address(i)=="84DD20C50B29" | list.Address(i)=="544A165E18AB" |
list.Address(i)=="20C38FD12605")
b(j)=ble(list.Address(i));
j=j+1;
end
end
Ng=j-1; % Number of Keyfobb devices
```

```
disp(['Number of Keyfob devices: ' num2str(Ng)])
if Ng==0
disp('No devices')
return
end
s_enable=struct; %initialize the structure
tab='abcdef'; %the fields of the structure
for i=1:Ng
s_enable.(tab(i))=characteristic(b(i),"FFA0","FFA1");
write( s_enable.(tab(i)),1);
end
s_x=struct; %initialize the structure
for i=1:Ng
s_x.(tab(i))=characteristic(b(i),"FFA0","FFA7");
s_x.(tab(i)).DataAvailableFcn = eval(['@displayCharacteristicData_temp'
num2str(i)]);
end
h=figure(1);
while (ishandle(h))
for i=1:Ng
var =eval(['V_temp' num2str(i)]);
timp =eval(['V_time' num2str(i)]);
subplot(Ng,1,i);plot(timp,var,'*');grid;title(['sensor' num2str(i)]);ylim([0
30])
xlabel('time')
ylabel('temperature, C')
drawnow;
end
pause(3)
end
clear b
%One of the three callback functions:
function displayCharacteristicData_temp1(src,evt)
global V_temp1
global V_time1
```

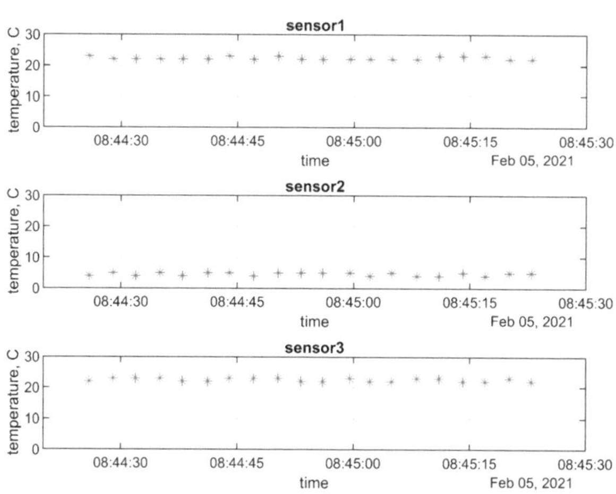

Figure 5.
The temperatures from the three sensors.

```
global N
[temp1] = read(src,'oldest');
time1=datetime(datestr(now,'HH:MM:SS.FFF'));
%update the last N=20 samples of temperature and time
V_temp1(1:N-1)=V_temp1(2:N);
V_temp1(N)=temp1;
V_time1(1:N-1)=V_time1(2:N);
V_time1(N)=time1;
end
```

The program runs in an endless loop and displays the last 20 values of the three temperatures in a MATLAB figure as in **Figure 5**. To stop the program, simply close the figure. In addition, the current date and time is displayed on the figure. One of the CC2541 Keyfob was on the outside sill of the window and therefore the resulting temperature was about 5 degree Celsius.

3. Using the accelerometer of CC2541Keyfob

This section presents a MATLAB application that accesses the accelerometer of the CC2541 Keyfob to read the 8-bit accelerations corresponding to the three axes. The program is similar to that of the previous section. There are also three callback functions, one for each axes. The period of reading data is set to 100 ms.

```
%%%%%%%%Display in real-time the last samples of three accelerations
%%%%%%%%of each axis %%%%%%%%%%%%%%%%%%%
clear all; close all;
b=ble("84DD20C50B29"); %BLE connection with 84DD20C50B29
enable=characteristic(b,"FFA0","FFA1")
write(enable,1) %enable accelerometer
x=characteristic(b,"FFA0","FFA3")
y=characteristic(b,"FFA0","FFA4")
z=characteristic(b,"FFA0","FFA5")
per=characteristic(b,"FFA0","FFA6")
read_per=read(per);
write(per,10);%establish the reading period to 10*10ms
read_per1=read(per);
global vector_x;global vector_y;global vector_z
global N; %number of samples
N=200;
vector_x=zeros(1,N;vector_y=zeros(1,N);vector_z=zeros(1,N);
axa =1:N;
x.DataAvailableFcn = @displayCharacteristicData_x; %functii callback,
they are %executed when sensor data are available
y.DataAvailableFcn = @displayCharacteristicData_y;
z.DataAvailableFcn = @displayCharacteristicData_z;
h=figure(1); %create the figure
while(ishandle(h)) %while the figure does exist data is displayed plot(axa,
vector_x*19.62/128,axa,vector_y*19.62/128,axa,vector_z*19.62/128);grid;leg-
end('ax','ay','az')
ylabel('accel., m/s2')
xlabel(' The most recent samples')
drawnow;
```

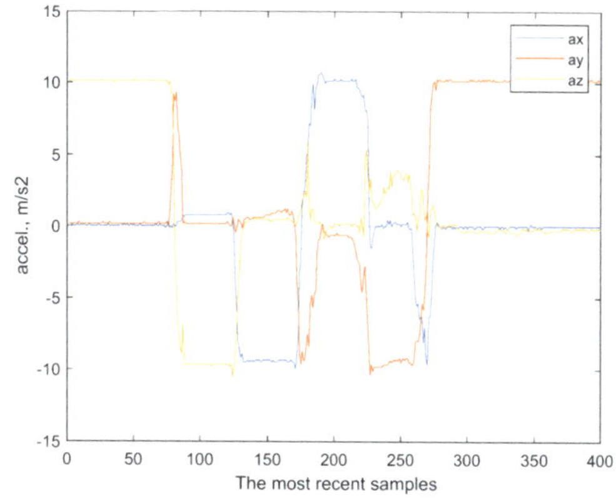

Figure 6.
The variation of the three accelerations.

```
end
clear b %clear the variable that represents the BLE connection or disconnect
%One of the three callback functions:
function displayCharacteristicData_x(src,evt)
global vector_x
global N
[data,timestamp] = read(src,'oldest');
fprintf('Time1 %s\n', datestr(now,'HH:MM:SS.FFF'))
if data>128
data=data-256;
end
%update the last N=200 samples of acceleration
vector_x(1:N-1)=vector_x(2:N);
vector_x(N)=data;
end
```

The program runs in an endless loop and displays the last N=200 samples of each the three axes. **Figure 6** presents a screenshot during the running of the program. During this time CC2541 Keyfob was moved such as one of the three axes was on the direction of gravitational force. Thus, most of the time one of the three axes has the absolute value close to g=9.81 m/s^2 while the other two are close to zero.

4. Using the movement sensor of CC2650 Sensor Tag

This section presents a MATLAB application that accesses the movement sensor of the device called CC2650 Sensor Tag. This movement sensor contains an accelerometer, a gyroscope and a magnetometer. If the accelerometer of the CC2541 Keyfob which was presented in the third section generates 8-bit data, all of the three sensors of CC2650 Sensor Tag generates 16-bit data.

The gyroscope is a three axis sensor that measures the angular rate, $\omega(t)$. By integrating the angular rate, the angular position $\alpha(t)$ is obtained as

$$\alpha(t) = \int \omega(t)dt \tag{1}$$

Thus, Eq. (1) can be implemented by trapezoidal method by using samples of $\omega(t)$ by

$$\alpha(t) = \alpha(t-1) + \frac{\omega(t) - \omega(t-1)}{2} dt \tag{2}$$

where dt is reciprocal of sample rate.

This angle is considered in comparison with the initial position of the gyroscope which is unknown. Using the integration, it generates an error because the gyroscope has an offset. That means its output is different to zero when the gyroscope is still. Thus, by integration it follows that the angle is changed. Therefore this offset must be removed [11].

The magnetometer measures the magnetic field. Thus, if there are no other fields, it measures the magnetic field of the earth. When the magnetometer is placed horizontally, it can measure the angle from the north, h, by

$$h = atan \frac{m_y}{m_x} \tag{3}$$

where m_x and m_y are its readings. Thus, both gyroscope and magnetometer can measure the same angle but the magnetometer has a reference which is the north. For this reason this angle is also called heading. Similar to the gyroscope, the magnetometer has a drawback too. Thus, its reading must be corrected by a process called calibration [12]. Basically, this implies a rotation of 360 degrees around its z axis in both senses followed by computation of calibrated data m_{xcal} and m_{ycal},

$$m_{xcal} = X_g \cdot m_x - X_{off} \tag{4}$$

and

$$m_{ycal} = Y_g \cdot m_y - Y_{off} \tag{5}$$

where

$$X_g = \frac{max\left\{\left(m_{x\ max} - m_{x\ min}\right), \left(m_{y\ max} - m_{y\ min}\right)\right\}}{m_{x\ max} - m_{x\ min}} \tag{6}$$

and

$$X_{off} = X_g\left(\frac{m_{x\ max} + m_{x\ min}}{2}\right) \tag{7}$$

while Y_g and Y_{off} have similar expressions. Also the magnetometer is very sensitive to the magnetic perturbations that can be generated by other materials from its proximity.

Regarding BLE, CC2650 Sensor Tag offers more services. The service that allows accessing the accelerometer, the gyroscope and the magnetometer has three characteristics, as shown in **Table 2**, where also the UUID can be seen. The first one is

Service Name	UUID	Char Name	UUID
Movement	F000AA80	Data	F000AA81
		Configure sensors	F000AA82
		Period	F000AA83

Table 2.
UUID for the movement service and its characteristics.

used to read data. The second one is used to enable the sensors. Each axis of the gyroscope and accelerometer can individually be enabled while the magnetometer can be enabled only for all axes. The third characteristic allows to establish the period of data reading. The data is presented as an 18 bytes string, where each sensor has a field of 6 bytes, two bytes for each axis, in the order: gyroscope, accelerometer and magnetometer. Actually UUID contain more digits but only the different part is presented in **Table 2** [8].

Because in this case the MATLAB programs are much longer than previous ones only some parts of the achieved programs are presented. Mainly, such a program has three parts:

- the first part when the sensor is still for a time while gyroscope data are gathered to compensate its offset; generally in this part 200 samples are acquired;
- the second part when the sensor is rotated with 360 degrees in both senses around z axis while the magnetometer data are gathered to compute the calibration; in this part also 200 samples are acquired;
- the last part has an indefinite duration when the sensor is moved while real-time data are displayed on the different figures.

Two programs are achieved, depending of the content of the last part. In this case there is a script and only a callback function. The period of data reading is 200 ms.

The first program computes and displays basic results from gyroscope and magnetometer.

Mainly it computes:

```
om=[om_x om_y om_z]; %%the vector with gyroscope readings
A=[a_x a_y a_z]; %%the vector with accelerometer readings
M=[m_x m_y m_z]; %%the vector with magnetometer readings
ang_z_g %the angular position (angle) around z axis, by gyroscope
ang_z_m %the angular position (angle) around z axis, by magnetometer
```

All the time the last N=200 samples of these measurements are available.

A part of the script is presented in the following.

```
clear;
close all;
global i
i=0; %%this index is used inside the callback function
global ang_z_g; global om_z_prev; global ang_z_m
global offset_x; global offset_y; global offset_z;
ang_z_g=0; om_z_prev=0; ang_z_m=0
offset_x=0; offset_y=0; offset_z=0 ;
bb=ble("B0B448BD7E05");% the address of CC2650 Sensor Tag
%%%%the scaling constants, according to the range
sca_a=8*9.81/32768; %%the accelerometer range is +-8g
sca_m=4800/32768; %%the magnetometer range is +-4800 uT
sca_om=250/32768; %%the gyroscope range is +-250 deg/s
conf=characteristic(bb, "F000AA80-0451-4000-B000-000000000000" ,
"F000AA82-0451-4000-B000-000000000000" );
write(conf,[127 02]) %%enable all 9 axes, 01 111 111
```

```
per=characteristic(bb, "F000AA80-0451-4000-B000-000000000000" ,
"F000AA83-0451-4000-B000-000000000000" );
coef=20;T=10*coef*0.001; %%period of reading in seconds
write(per,coef) %%coef can be minimum 10
data=characteristic(bb, "F000AA80-0451-4000-B000-000000000000" ,
"F000AA81-0451-4000-B000-000000000000" );
data.DataAvailableFcn = @displayCharacteristicData_STag;
%A part of the callback function is presented in the following:
function displayCharacteristicData_STag(src,evt)
global i; global ang_z_g; global om_z_prev; global ang_z_m
global offset_x; global offset_y; global offset_z;
[data,timestamp] = read(src,'oldest'); %%read the current data
%%the gyroscope data
if data(2)>128
om_x=data(2)*256+data(1)-2^16;
else
om_x=data(2)*256+data(1);
end
if data(4)>128
om_y=data(4)*256+data(3)-2^16;
else
om_y=data(4)*256+data(3);
end
if data(6)>128
om_z=data(6)*256+data(5)-2^16;
else
om_z=data(6)*256+data(5);
end
om_x=om_x*sca_om;
om_y=om_y*sca_om;
om_z=om_z*sca_om;
%%correct the gyroscope's offset
if i<=200
offset_x=offset_x+om_x;
offset_y=offset_y+om_y;
offset_z=offset_z+om_z;
i=i+1;
end
om_x=om_x-offset_x/200;
om_y=om_y-offset_y/200;
om_z=om_z-offset_z/200;
om=[om_x om_y om_z]; %%the vector with gyroscope readings
if i>200
%compute the current angle by integration
ang_z_g=ang_z_g+(om_z+om_z_prev)*T/2; %T is reading period
om_z_prev=om_z; %previous value becomes current value
%%the accelerometer data
if data(8)>128
a_x=data(8)*256+data(7)-2^16;
else
a_x=data(8)*256+data(7);
end
if data(10)>128
```

DOI: http://dx.doi.org/10.5772/intechopen.95814

```
a_y=data(10)*256+data(9)-2^16;
else
a_y=data(10)*256+data(9);
end
if data(12)>128
a_z=data(12)*256+data(11)-2^16;
else
a_z=data(12)*256+data(11);
end
a_x=a_x*sca_a;
a_y=a_y*sca_a;
a_z=a_z*sca_a;
A=[a_x a_y a_z]; %%the vector with accelerometer readings
%%%%%the magnetometer data
if data(14)>128
m_x=data(14)*256+data(13)-2^16;
else
m_x=data(14)*256+data(13);
end
if data(16)>128
m_y=data(16)*256+data(15)-2^16;
else
m_y=data(16)*256+data(15);
end
if data(18)>128
m_z=data(18)*256+data(17)-2^16;
else
m_z=data(18)*256+data(17);
end
m_x=m_x*sca_m;
m_y=m_y*sca_m;
m_z=m_z*sca_m;
M=[m_x m_y m_z]; %%the vector with magnetometer readings
if i>400
%%compute the heading angle using magnetometer
AA= m_y*Ysf+Yoff; %Ysf and Yoff are constants for calibration
BB= m_x*Xsf+Xoff; %Xsf and Xoff are constants for calibration
%the part of the program that computes the above constants is not %
presented
ang_z_m=180/pi*atan2(AA,BB) %%the magnetometer-based heading %angle
end
end %%end if i>200
```

By running the previous program, the last 200 samples of some of the measurements obtained from the gyroscope and magnetometer are displayed in real-time.

Thus, the two waveforms of the top of **Figure 7** are achieved using the gyroscope while the other two from the bottom part are achieved using the magnetometer. In each case the heading is computed. The gyroscope-based angle around z axis or heading is computed using the angular rate around the z axis, see the second waveform. The heading computed by the magnetometer presented in the third waveform is based on its x and y reading which are presented in the last waveform. It can be seen that the two waveforms that represent the heading have the same variation, except at

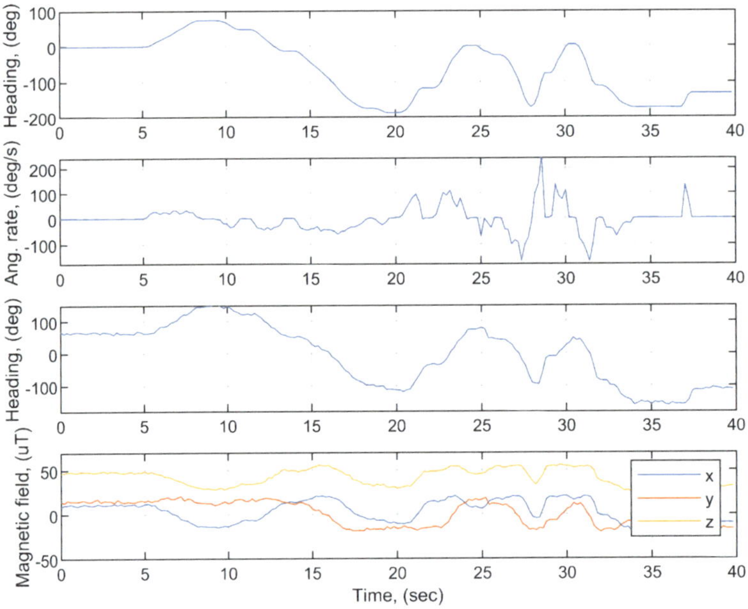

Figure 7.
Some measurements obtained from gyroscope (the two waveforms at the top) and magnetometer (the two waveforms at the bottom).

the start. Thus the gyroscope-based heading starts from zero while magnetometer-based heading starts from about 60 degrees because it indicates the north.

By using the movement sensors, 3D orientation of a device can be computed [13–17]. This can be represented in three ways: quaternion, Direction Cosine Matrix (DCM) and Euler angles. The last representation means the rotational angles around the three axes, called pitch, roll and yaw but has a disadvantage because can reach in a singularity state. DCM does not have a singularity state but needs 3x3 elements. Thus the best representation is quaternion which represents a complex number having four components [13],

$$q = \left[cos\frac{\alpha}{2} \;\; e_x sin\frac{\alpha}{2} \;\; e_y sin\frac{\alpha}{2} \;\; e_z \; sin\frac{\alpha}{2} \right] \tag{8}$$

where α is the rotation angle and e represents the rotation axis.

Using the accelerometer allows only computing pitch and roll angles because a rotation around z axis does not change any of the three outputs. Thus the four elements of the quaternion, denoted q_{acc} can be computed by [13]

$$q_{acc} = \begin{cases} \left[\sqrt{\frac{a_z+1}{2}} \;\; -\frac{a_y}{\sqrt{2(a_z+1)}} \;\; \frac{a_x}{\sqrt{2(a_z+1)}} \;\; 0 \right], & a_z \geq 0 \\ \left[-\frac{a_y}{\sqrt{2(1-a_z)}} \;\; \sqrt{\frac{1-a_z}{2}} \;\; 0 \;\; \frac{a_x}{\sqrt{2(1-a_z)}} \right], & a_z < 0 \end{cases} \tag{9}$$

Eq. (9) can be very easily implemented in MATLAB and then the quaternion can be generated by the function *quaternion*(). However, in order to compute the third angle (yaw) the magnetometer or gyroscope readings are necessary and more difficult equations are generated [13]. Thus, a better solution is using the functions from Sensor Fusion and Tracking Toolbox of MATLAB [18] that allows estimation of 3D orientation. The function *imufilter*() uses only accelerometer and gyroscope while the function *complementaryFilter*() uses all the three sensors. Thus the **Figure 8** presents in the top panel the angles obtained by MATLAB implementation of Eq. (9) while in the bottom panel are presented the results obtained using the *imufilter*() function. It can be seen that the waveforms are very similar and also the angle around z axis is zero for the first or close to zero for the second as expected.

Both *imufilter*() and *complementaryFilter*() return the result as an object. Then the object can be called as a function having the sensor measurements as arguments and returns a quaternion. If this quaternion becomes the argument of the *viewer*() function [16] it follows the display of a cube that moves in real-time that imitates the moving of the CC2650 Sensor Tag, as in **Figure 9**. However, two files,

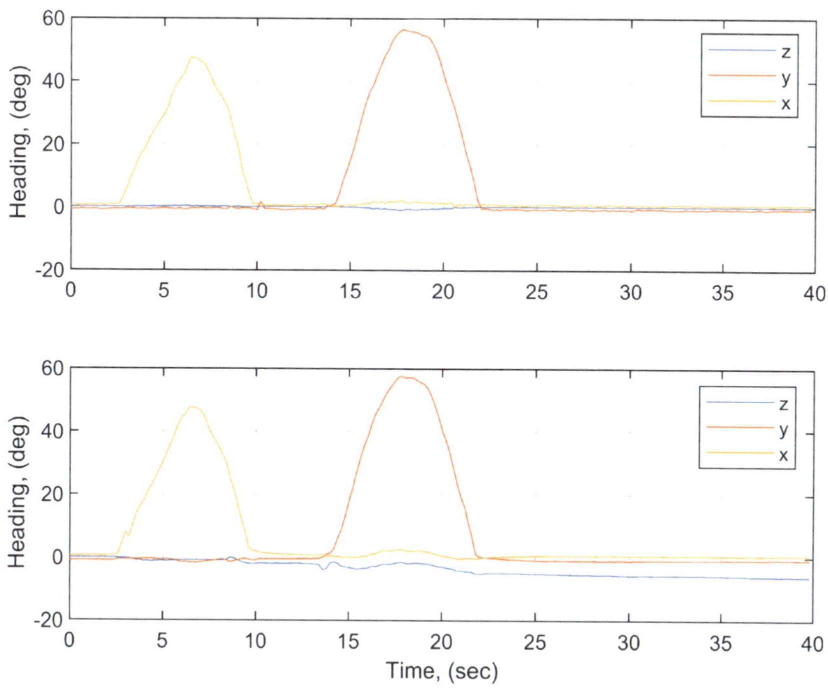

Figure 8.
The Euler angles computed by Eq. (9), top, and by imufilter(), bottom; the sensor was moved around x and y axis.

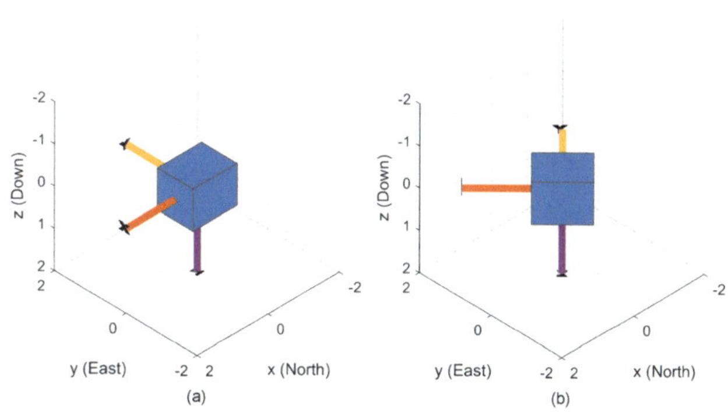

Figure 9.
The cube that imitates the moving of the CC2650 Sensor Tag, the initial position-left, the position after a 45 degrees rotation around z axis-right.

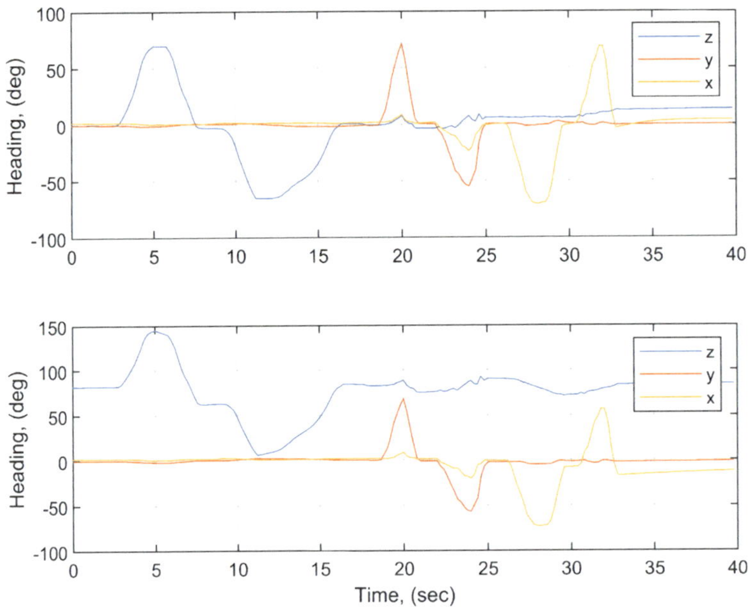

Figure 10.
The Euler angles computed by imufilter*(), top, and by* complementaryFilter*(), bottom; the sensor was moved around all the three axes.*

HelperOrientationViewer.m and HelperBox.m, must be in the current folder in order to use the function *viewer*() [18].

On the other hand, using the function *eulerd*(), the quaternion form can be converted in pitch, roll and yaw angles (Euler angles) that can be represented as waveforms. This method was used to obtain the results in **Figure 8**. Now **Figure 10** presents the three angles computed using *imufilter*(), in the top part and *complementaryFilter*(), in the bottom part. It can be seen that the angles have the same variation except that the heading around the z axis starts with zero at the top and by about 80 degrees in the bottom. This is because the *complementaryFilter*() uses the magnetometer and thus indicates the angle with respect to the north.

The program that implements these facilities is very similar to the previous one. In the following sections, only the new elements are presented.

First the new elements of the script are presented.

```
viewer = HelperOrientationViewer;
SRate=1/T;
ifilt_imu = imufilter('SampleRate', SRate);
ifilt_com = complementaryFilter('SampleRate', SRate);
h=figure(1);
while(ishandle(h))
plot(v_om');grid;legend('x','y','z') %%display the gyroscope readings
drawnow;
if i>200
viewer(qahrs_imu); %%imu quaternions are used to move the cube
end
end

%Then it follows the new elements of the callback function.
%%%%%%%%%%%%%%%%%%%quaternion%%%%%%%%%%
```

```
qahrs_imu=ifilt_imu(A,om*pi/180); %% for imufilter
qahrs_com=ifilt_com(A,om*pi/180,M); %%for complementaryFilter
eulerAnglesDegrees_imu=eulerd(qahrs_imu, 'ZYX','frame')
eulerAnglesDegrees_com=eulerd(qahrs_com, 'ZYX','frame')
```

5. Conclusions

The main MATLAB contributions of this chapter are:

- using the new introduced MATLAB functions to access BLE devices and to implement a BLE sensors network
- using the table type MATLAB to check if the desired sensors are among the discovered BLE devices
- using the structure type MATLAB having a variable number of fields to handle the discovered number of BLE devices
- retaining and updating the most recent samples of different measurements corresponding to BLE sensors and display them in real-time
- using the quaternions to handle the 3D orientation of an object
- using the new introduced MATLAB functions from Sensor Fusion and Tracking Toolbox to determine the parameters that describes 3D orientation
- displaying the cube that imitates in real-time the moving of CC2650 Sensor Tag
- as a future work, the MATLAB can be used to estimate the position of an object along with 3D orientation; in this way the tracking of an object can be completed.

Author details

Septimiu Mischie
Politehnica University Timisoara, Timisoara, Romania

*Address all correspondence to: septimiu.mischie@upt.ro

References

[1] Katsikis V, editor. MATLAB. A fundamental tool for scientific computing and engineering applications. Volume1. IntechOpen; 2012. DOI: 10.5772/2557, Volume 2. IntechOpen; 2012. DOI: 10.5772/3338, Volume 3 IntechOpen; 2012. DOI: 10.5772/3339

[2] Mohamad Omar Al Kalaa, Refai H. Bluetooth Standard v4.1: Simulation the Bluetooth Low Energy Data Channel Selection Algorithm. In: Proceedings of Globecom 2014 Workshop – Telecommunication Standards – From Research to Standards. P. 729-733

[3] Afaneh M. Bluetooth5 & Bluetooth Low Energy. A Developer's Guide, e-book, 2018. Available from: https://www.novelbits.io/bluetooth-5-developers-e-book/ [Accessed: 2020-12-02]

[4] Bluetooth. Available from https://www.bluetooth.com/ [Accessed: 2020-12-02]

[5] Wu Taiyang, Redoute J.M, Yuce M. A wireless Implantable Sensor Design with Subcutaneous Energy Harvesting for Long-Term IoT Healthcare Applications. In IEEE Access, vol. 6, 2018, p.35801-35808.

[6] MATLAB Bluetooth Low Energy. Available from https://www.mathworks.com/help/matlab/bluetooth-low-energy-communication.html [Accessed: 2020-12-02]

[7] BLE Device Monitor. Available from https://processors.wiki.ti.com/index.php/BLE_Device_Monitor_User_Guide [Accessed: 2020-12-02]

[8] CC2650 Sensor Tag. Available from https://processors.wiki.ti.com/index.php/CC2650_SensorTag_User%27s_Guide [Accessed: 2020-12-02]

[9] CC2541 Keyfob, Available from https://www.ti.com/tool/CC2541KEYFOB-RD [Accessed: 2020-12-02]

[10] Mischie S. On the Development of Bluetooth Low Energy Devices. In: Proceedings of COMM 2018, Bucharest. p.339-344

[11] Mischie S. A MATLAB Graphical Interface to evaluate CC2650 Sensor Tag. In: Proceedings of 22nd IMEKO-TC4 International Symposium, Iasi, Romania, 2017

[12] Fang J, Sun H, Zhang X, Tao Y. A novel Calibration Method of Magnetic Compass Based on Ellipsoid Fitting. IEEE Trans. on Instrumentation and Measurement, vol. 60, no.6, June 2011, p. 2053-2061

[13] Valenti R, Dryanovski I, Xiao J. A Linear Kalman Filter for MARG Orientation Estimation Using the Algebraic Quaternion Algorithm. In IEEE Trans. On Instrumentation and Measurement, vol. 65, no.2 2016, p. 467-481, DOI:10.1109/TIM.2015.2498998

[14] Yadav R.H, Bhattarai B., Gang H.S. Pyiun J.Y. Trusted K Nearest Batesian Estimation for Indoor Positioning System. In IEEE Access, vo. 7 2019, p.51484-51498. DOI: 10.1109/ACCESS.2019.2910314

[15] Manos A., Kleian I., Hazan T. Gravity-Based Methods for Heading Computation in Pedestrian Dead Reckoning. In Sensors 2019, 19(5), 1170. DOI: 10.3390/s19051170

[16] Li G., Geng E. Ye Z., Xu Y., Lin J., Pang Y. Indoor Positioning Algorithm Based on the Improved RSSI Distance Model, In Sensors 2018, 18(9), 2820. DOI:10.3390/s18092820

[17] Thomas C, editor. Sensor Fusion. Foundation and Applications. IntechOpen 2011. DOI: 10.5772/680

[18] MATLAB. Sensor Fusion. Tracking Toolbox, 2020. Available from https://www.mathworks.com/products/sensor-fusion-and-tracking.html [Accessed: 2020-12-02]

Section 5

Fuzzy Control of Electric Drives

Chapter 5

Matlab Program Library for Modeling and Simulating Control Systems for Electric Drives Based on Fuzzy Logic

Constantin Volosencu

Abstract

Fuzzy control of the speed of electric drives is an alternative in the field of the control system. Modeling and simulation of electric drive control systems based on fuzzy logic is an important step in design and development. This chapter provides a complete means of modeling and simulation of fuzzy control systems for DC motors, induction motors, and permanent magnet synchronous motors, made in the Matlab/Simulink program environment, useful for performing complex analyzes. The functioning of the programs is demonstrated by an example of characteristics obtained practically, with a functioning regime often encountered in practice.

Keywords: simulation, modeling, control of electric drives, DC motors, induction motors, permanent magnet synchronous motors, fuzzy PI controllers

1. Introduction

Electric drives play an important role in the development of machine tools, production systems, means of transport, and many other practical applications. The purpose of using electric drive control systems is to ensure good performance indicators. The use of fuzzy logic in the control of electric drives ensures the realization of high-performance systems. Modeling and simulation of electric drive control systems based on fuzzy logic are an important means in their design. This chapter presents a library of Matlab/Simulink programs designed to model and simulate electrical drive control systems based on fuzzy speed PI controllers.

In the literature, the control of electric drives based on fuzzy logic is studied in many works. Several Matlab programs for modeling and simulating electric drive systems based on fuzzy logic are presented on the Matlab website as well. Programs for modeling and simulating fuzzy DC drive systems are presented in [1–3]. Programs for modeling and simulating fuzzy driving systems of induction motors are presented in [4, 5]. The Matlab software has the facilities for modeling and simulating fuzzy systems [6] and electric drive systems [7, 8]. The problem of intelligent control of electric drives has been addressed in numerous papers over the years, including application of expert systems, fuzzy logic and neural networks in electric drives [9], fuzzy control of switched reluctance motor drives [10], or fuzzy adaptive vector control of induction motor drives [11]. The basic management systems of electric actuators have been treated in numerous works in the literature as well. The problem of using electric machines in variable speed control systems is treated in [12], principles of motion control with induction motors are presented in [13] and with permanent magnet, AC machines in [14], the issues of pulse-width modulation for electronic power conversion are presented in [15], principles of modeling and simulation of electric drive control systems are presented in [16]. The basic design of fuzzy PID controllers is shown in [17].

The author of this chapter published the results of his research in the field of fuzzy control of electric drives in specialized literature, addressing the following issues: speed control based on fuzzy PI controllers of DC machines [18], of synchronous machines with permanent magnets [19], of AC machines in general [20], demonstration of the robustness of fuzzy control systems of electric machines [21], analysis of the basic properties of fuzzy control systems [22, 23], tuning of fuzzy PID regulators [24, 25] and analysis of the stability of fuzzy control systems [26].

The chapter presents program libraries dedicated to fuzzy speed regulation of the main electric motors used in practice: direct current motors, induction motors, and permanent magnet synchronous motors, respectively in subsections 2, 3, and 4. For each element of the control systems are presented the equations used in modeling and the related subprograms. For each control system, the transient regime characteristics obtained by simulation are presented. Based on the transient regime characteristics, the values obtained for the performance indicators of the control systems are highlighted, such as overshoot, rise time, error, and others.

2. Library programs

2.1 DC motor

The developed programs solve the problem of speed control of the DC machine with the control system from **Figure 1**.

The speed control structure from **Figure 1** has the following components: MCC - DC motor, ML - load machine, CONV - power converter, RG-i - current controller, RF-Ω - fuzzy speed controller, Ti - current sensor, TΩ - speed sensor, CAN, CNA - analog to digital and digital to analog converters, MM-ISI - DC motor with state-space equations, Lim - anti-wind-up circuit. The control system variables are: Ω^* - speed reference, Ω - motor speed, Ω_m - measured speed, M - motor torque, M_s - load torque, e_Ω - speed error, i^* - current reference, i_{am} - measured current, e_i - current error, u_i - command voltage, u_a - motor armature voltage, i_a - current motor, u_e - excitation voltage.

The Simulink block diagram of the speed control system of DC drives based on the fuzzy PI controller is presented in **Figure 2**.

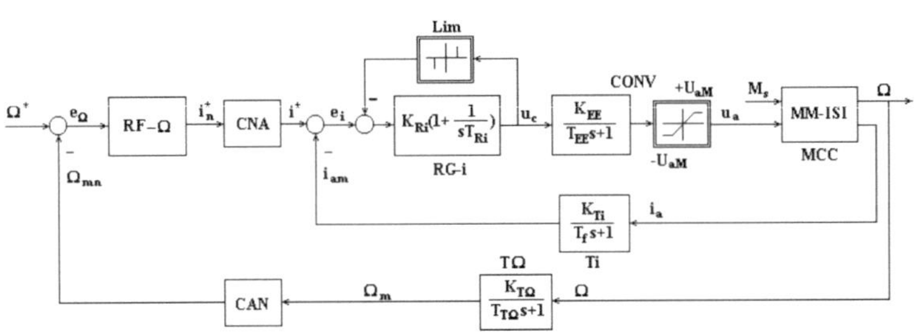

Figure 1.
Block diagram of the DC motor fuzzy speed control system structure.

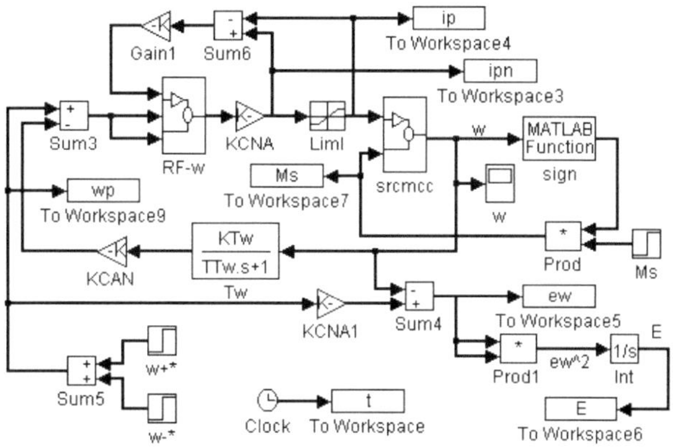

Figure 2.
Simulink block diagram of the fuzzy control system of DC drive.

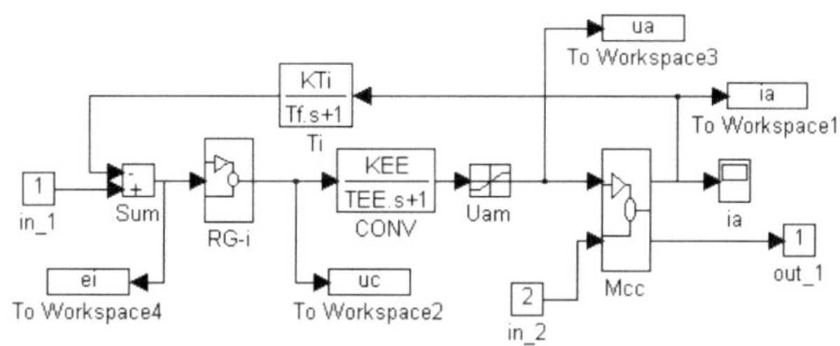

Figure 3.
Simulink diagram of current control loop srcmcc.

The block *RF-w* in **Figure 1** represents the Simulink model of the fuzzy speed controller. The other part of the Simulink diagram is a conventional control part of a linear control system.

The block *srcmcc* is the internal current control loop and it has the Simulink block diagram from **Figure 3**.

The block *Mcc* represents the DC motor and it has the equations:

$$\begin{aligned} \frac{di_a}{dt} &= -\frac{R_a}{L_a}i_a - \frac{k_e}{L_a}\Omega + \frac{1}{L_a}u_a \\ \frac{d\Omega}{dt} &= \frac{k_m}{J}i_a - \frac{k_f}{J}\Omega - \frac{1}{J}M_s \end{aligned} \tag{1}$$

The power converter has the block *CONV* and the current sensor has the block *Ti*. Their transfer functions are presented in their blocks. The current controller has a PI linear transfer function. The armature voltage u_a is limited. The speed sensor has the block *Tw*. The current controller has anti-windup protection. The current reference is limited. The simulation diagram allows simulations in four quadrants for speed and torque. The diagram calculates a quadratic performance criterion of speed error with the formula:

$$E = \int e^2 dt \tag{2}$$

A theoretic design for fuzzy speed control systems for DC drives is presented in [27].

2.2 Induction motor

The structure of the fuzzy control system of the induction machine is presented in **Figure 4**.

The meanings of the notations in the **Figure 4** are as follows: MAS - induction machine; ML- working machine; CONV- power electronic convertor; RG-i - stator phase current controllers; Ti—current sensor; Tθ—position sensor; TΩ—speed sensor; TCI—inverse Park coordinate transformation; BCC—block for calculating the reference currents in the vector control structure of the asynchronous machine with rotor flux orientation; RF-Ω—speed fuzzy PI controller; Ω—rotor speed; Ω*—speed reference; M^*—torque reference; $\phi_r{}^*$—rotor flux reference; $\theta_s{}^*$—stator phasor position reference; $i_d{}^*$, $i_q{}^*$—d, q stator current references; $i_s{}^*$—stator current reference; Ω_m—measured speed; θ_m—measured rotor position; i_{sm}—measured stator currents; e_Ω—speed error; e_i—current error; S_i—control signals for power converter switches with pulse width modulation; V_{cc}—DC converter supply voltage; u_s—stator voltages; M_s—load torque. The induction motor is vector controlled with rotor flux orientation, with rotor flux reference ϕ^*_r and torque reference M^* and the measured rotor position θ_m. The reference stator currents are calculated with the block TCI. The current regulators RG-i give the pulse width modulation signals S_i for the electronic power converter CONV, fed from a DC voltage source V_{CC}. The power converter CONV gives the stator voltages u_s.

The speed control system of induction motors based on the fuzzy PI controller is presented in **Figure 5**.

The model from **Figure 5** implements an induction motor control structure with indirect field orientation in rotor coordinates [13, 28]. The block *mas* represents the induction motor and it has the Equations [13, 29, 30]:

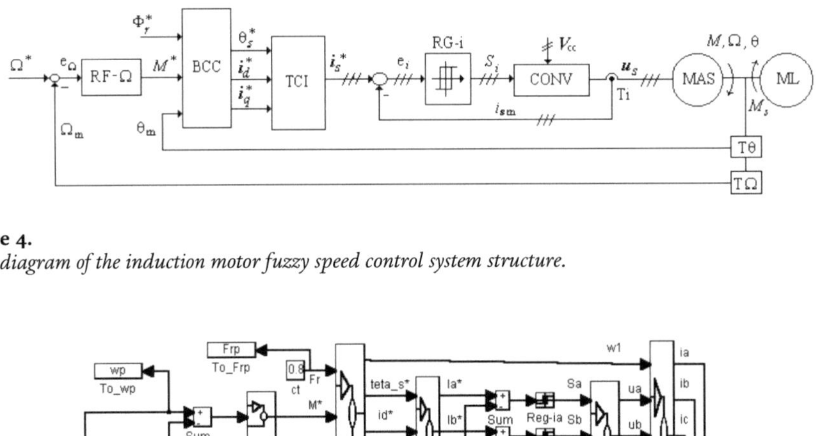

Figure 4.
Block diagram of the induction motor fuzzy speed control system structure.

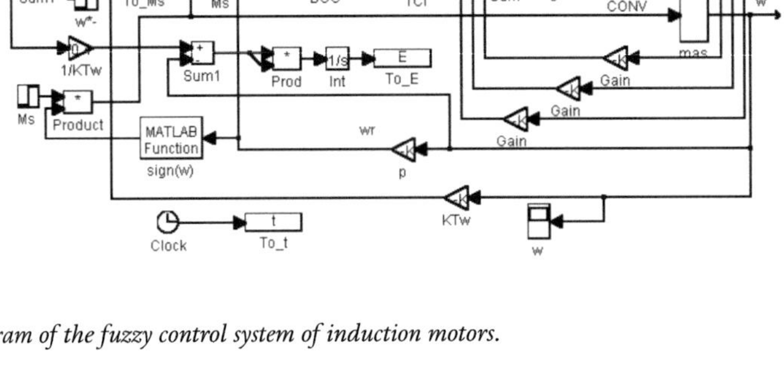

Figure 5.
Simulink diagram of the fuzzy control system of induction motors.

$$\underline{u}_{sf} = \underline{u}_s e^{-j\theta_f}, \underline{i}_{sf} = \underline{i}_s e^{-j\theta_f}, \underline{\Phi}_{sf} = \underline{\Phi}_s e^{-j\theta_f}, \underline{u}_{rf} = \underline{u}_r e^{-j\theta_f+\theta}, \underline{i}_{rf} = \underline{i}_r e^{-j\theta_f+\theta},$$

$$\underline{\Phi}_{rf} = \underline{\Phi}_r e^{-j\theta_f+\theta}, \underline{u}_{sf} = R_s \underline{i}_{sf} + \frac{d\underline{\Phi}_{sf}}{dt} + j\omega_f \underline{\Phi}_{sf},$$

$$\underline{u}_{rf} = R_r \underline{i}_{rf} + \frac{d\underline{\Phi}_{rf}}{dt} + j(\omega_f - \omega)\underline{\Phi}_{rf}, M = \frac{3}{2} n_p L_m \operatorname{Re}\left\{ j\underline{i}_{sf} \cdot \underline{i}_{rf} \right\}, J\frac{d\Omega}{dt} = M - k_f\Omega - M_s \tag{3}$$

The Eqs. (3) are written for the general flux, and in the case of vector control with rotor flux orientation, the flux is the rotor flux: $\theta_f = \theta_r$. CONV is a power inverter functioning in comutation, TCI is the inverse Park coordinate transformation [30], BCC is for calculating the reference currents in the vector control structure of the asynchronous machine with rotor flux orientation. BCC has the block diagram from **Figure 6** [13, 30].

Where n_p - the pole pair number, $L_{m,r}$ - magnetic and rotor inductances, s - slip frequency, R_r - rotor resistance, ω_s^* - stator frequency reference. Two-position current controllers RG-i with hysteresis are used.

2.3 Permanent magnet synchronous motor

The structure of the fuzzy control system of the permanent magnet synchronous motor is presented in **Figure 7**.

The meanings of the notations in **Figure 7** are the same as in **Figure 4**, and MSMP is the permanent magnet synchronous machine. The permanent magnet synchronous motor is vector controlled with rotor flux orientation with rotor q current i^*_q as a torque reference, the rotor d current i^*_d at zero as a flux reference ϕ^*_r, and the measured rotor position θ_m.

The speed control system of induction motors based on the fuzzy PI controller is presented in **Figure 8**.

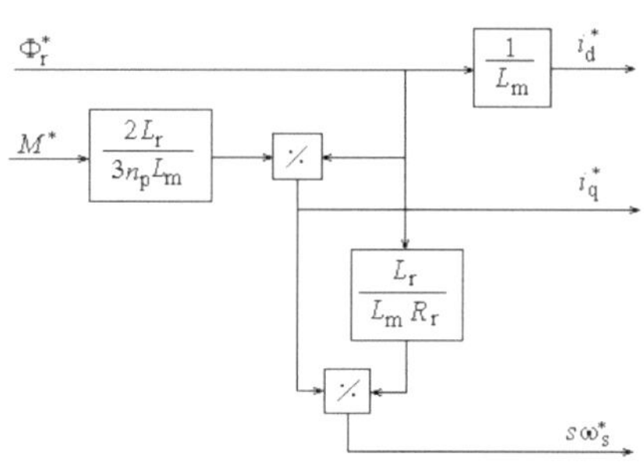

Figure 6.
The block for calculating the reference currents in the vector control structure of the asynchronous machine with rotor flux orientation (BCC).

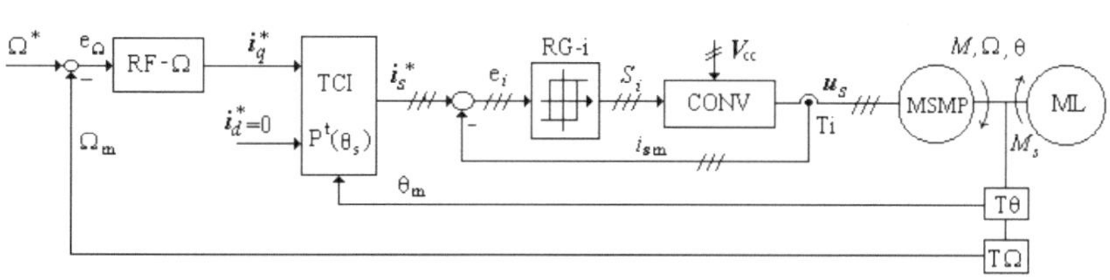

Figure 7.
Block diagram of the induction motor fuzzy speed control system structure.

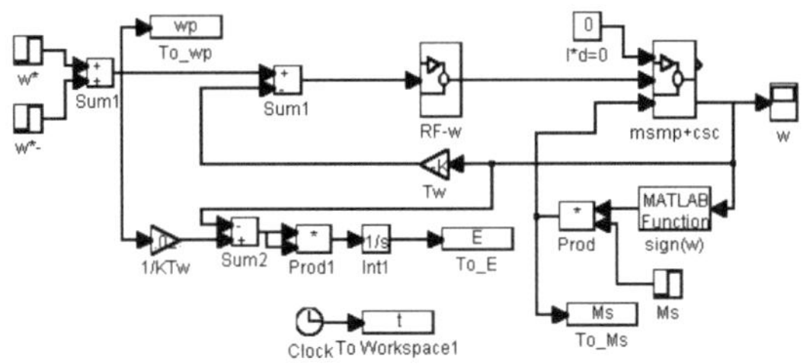

Figure 8.
Simulink diagram of the fuzzy control system of permanent magnet synchronous motors.

This diagram implements the rotor flux oriented control structure [14]. The block *msmp + csc* includes the permanent magnet synchronous motor, with Equations [14, 30]:

$$
\begin{aligned}
&\underline{u}_{sf} = \underline{u}_s e^{-j\theta}, \underline{i}_{sf} = \underline{i}_s e^{-j\theta}, \underline{\Phi}_{sf} = \underline{\Phi}_s e^{-j\theta}, \underline{u}_{sf} = R_s \underline{i}_{sf} + \frac{d\underline{\Phi}_{sf}}{dt} + j\omega\underline{\Phi}_{sf}, \\
&M = \frac{3}{2}\,\mathrm{Re}\left\{ j\underline{\Phi}_{sf}.\underline{i}_{sf} \right\}, J\frac{d\Omega}{dt} = M - k_f\Omega - M_s
\end{aligned}
\tag{4}
$$

The current controllers and the Park inverse coordinator transformation, like in the case of the induction motor.

2.4 Fuzzy speed controller

The Simulink diagram of fuzzy speed controller RF-w is presented in **Figure 9**. It is developed based on the block diagram of the fuzzy controller RG-F from **Figure 10**.

The fuzzy controller has Mamdani's structure with the fuzzification of the input variables x_i, inference of the fuzzy values x_f, with a rule base, and defuzzification of the fuzzy command u_f. The fuzzy block gives the command u_d. Different membership functions for the input and output variables, different inference methods, and different rule bases may be chosen [17, 19, 21, 22]. Here are some Matlab programs for fuzzy computing, as follows.

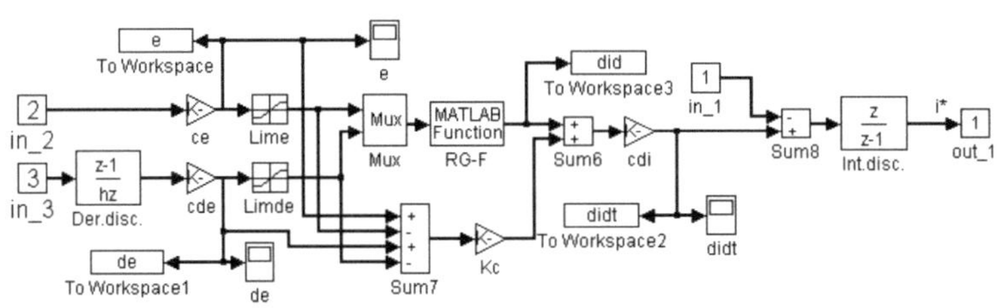

Figure 9.
Simulink diagram of fuzzy speed controller RF-w.

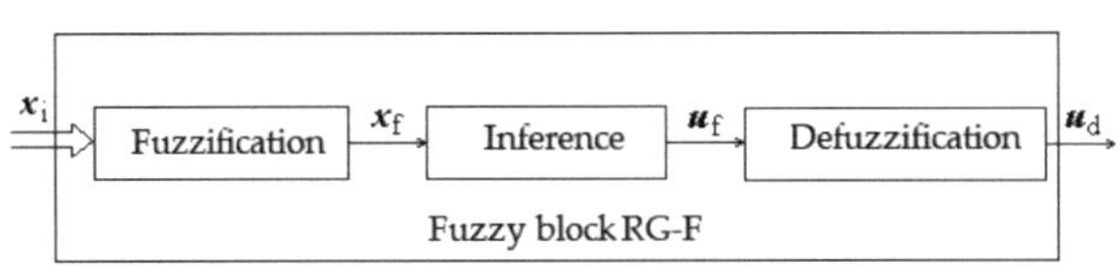

Figure 10.
The block diagram of the fuzzy controller RG-F.

A program for defining the function of triangle membership:

```
function [m]=triunghi(x,a,b,c)
% parameters: a<b<c / m(a)=m(c)=0, m(b)=1
n=length(x);
for i=1:n
    if x(i)<=a | x(i)>=c
        m(i)=0;
    elseif x(i)>a & x(i)<b
        m(i)=1/(b-a)*x(i)+a/(a-b);
    elseif x(i)==b
        m(i)=1;
    else
        m(i)=x(i)/(b-c)+c/(c-b);
end
end
```

A program for defining the function of decreasing trapezoidal membership:

```
function [m]=trapezd(x,a,b)
% parameters: a<b, m(a)=1, m(b)=0
n=length(x);
for i=1:n
    if x(i)<=a
        m(i)=1;
    elseif x(i)<b & x(i)>a
        m(i)=x(i)/(a-b)+b/(b-a);
    else
        m(i)=0;
end
end
```

A program for defining the function of increasing trapezoidal membership:

```
function [m]=trapezc(x,a,b)
% parameters: a<b, m(a)=0, m(b)=0
n=length(x);
for i=1:n
    if x(i)<=a
        m(i)=0;
    elseif x(i)>=b
        m(i)=1;
    else
        m(i)=x(i)/(b-a)+a/(a-b);
end
end
end
```

A program for calculating the rule base 3–3 for DC motors:

```
% Loading control system parameters:
load pudmcc
% Calculating the margins of discourse universes:
diN=IN/Ki/5/Ti;
diM=IM/Ki/5/Ti;
```

```
ewM=KCAN*KTw*2*wb;
ewb=ewM/2;
deMt=KCAN*KTw/J*(MMt+kf*wb);
deMc=KCAN*KTw/J*(MMc+kf*wb);
pdi=diM/20;
pe=ewM/20;
pde=deMt/20;
% discourse universes:
udi=[-diM:pdi:diM];
ue=[-ewM:pe:ewM];
ude=[-deMt:pde:deMt];
% Definition of fuzzy values:
NBdi=trapezd(udi,-diN,0);
ZEdi=triunghi(udi,-diN,0,diN);
PBdi=trapezc(udi,0,diN);
NBe=trapezd(ue,-ewb,0);
ZEe=triunghi(ue,-ewb,0,ewb);
PBe=trapezc(ue,0,ewb);
NBde=trapezd(ude,-deMc,0);
ZEde=triunghi(ude,-deMc,0,deMc);
PBde=trapezc(ude,0,deMc);
% Normalization of discourse universes:
ue=ue/ewb;
ude=ude/deMc;
udi=udi/diN;
subplot(3,1,1),plot(ue,NBe,ue,ZEe,ue,PBe);
     xlabel('ew');ylabel('me');grid
subplot(3,1,2),plot(ude,NBde,ude,ZEde,ude,PBde);
     xlabel('de');ylabel('mde');grid
subplot(3,1,3),plot(udi,NBdi,udi,ZEdi,udi,PBdi);
     xlabel('di');ylabel('mdi');grid
% Table of rules
A1=[NBe; ZEe; PBe;
   NBe; ZEe; PBe;
   NBe; ZEe; PBe];
A2=[NBde; NBde; NBde;
     ZEde; ZEde; ZEde;
     PBde; PBde; PBde];
B=[NBdi; NBdi; ZEdi;
    NBdi; ZEdi; PBdi;
    ZEdi; PBdi; PBdi];
save daterf3 ue ude udi A1 A2 B ewM deMt
```

A program for fuzzy block implementation in Simulink scheme:

```
function [dip]=bdf(e,de,A1,ue,A2,ude,B,udi)
% Input variables: error e, error deivative: de
% Output variable: dip
% Inference with Larsen max-min method:
dif=infermm(A1,e,ue,A2,de,ude,B);
% Defuzzification with centre of gravity method:
dip=defzfir(dif,udi,1);
end
```

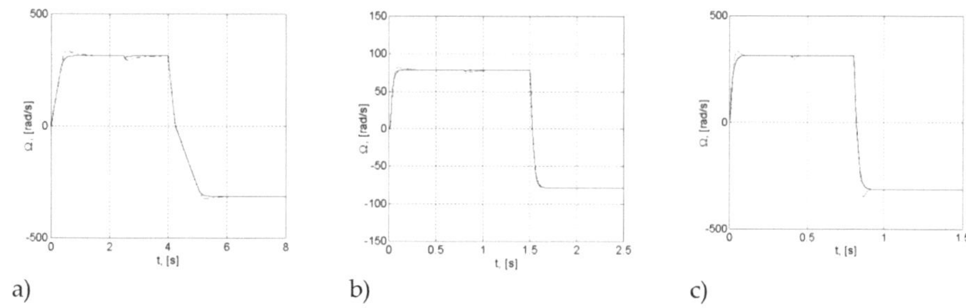

Figure 11.
Speed characteristics for direct current machine a), induction machine b) and synchronous machine with permanent magnets c), for fuzzy regulation with continuous line and for conventional regulation with a dashed line.

The fuzzy block RG-F has algebraic properties and the sector property presented in [22–24]. The fuzzy controller RF-w may be designed using a pseudo-equivalence with a linear PI controller with a grapho-analytical method [25–27], based on its input–output transfer characteristics [22–24].

3. Speed characteristics

With the help of the programs presented above, transient characteristics can be obtained for various operating regimes, which can be chosen by the signals applied to the speed prescription inputs and to the disturbing inputs of the load torques. Thus, transient regime characteristics can be obtained for speeds, currents, voltages, fluxes, mechanical torques, regulation errors, and others. These programs allow complex analyzes of the behavior of speed control systems based on fuzzy PI controllers. The fuzzy PI controller can be replaced with a conventional, linear PI controller. In order to demonstrate the good functioning of the programs, the following is an example of an operating regime often encountered in practice for the three-speed regulation structures. The simulated operating regime consists of: starting the machine idle up to the nominal speed, loading it with a nominal mechanical torque, and reversing the load. It was also chosen to exemplify the case when a conventional linear PI speed regulator is used. The characteristics in the two cases - fuzzy and linear - are presented in the same graph, and the same coordinate axes, for example. **Figure 11a–c** show the speed characteristics for the DC machine, induction machines, and the permanent magnet synchronous machine, respectively.

It is observed that in the case of fuzzy control better quality control indicators are obtained: zero overshoot, shorter rise time, shorter time for elimination of load torque effect, etc. [18–20]. The fuzzy control structures are global absolute internal stable and external BIBO stable [28]. The fuzzy control structures are robust at parameter identification errors and at the perturbation from the load torque [21].

4. Conclusion

The chapter presents a library of Matlab/Simulink programs for the control of electric drives. Thus, Simulink schemes are presented for modeling and simulating the fuzzy speed control systems of direct current machines, induction machines with vector control with rotor flux orientation, and synchronous machines with permanent magnets. Matlab/Simulink programs are presented for modeling and simulating fuzzy PI controllers based on the Mamdani structure. To demonstrate the operation of the programs, the characteristics of the speed obtained in the case of the three adjustment structures are presented. Fuzzy system modeling programs can be developed for various types of membership functions, inference methods, and rule bases.

Conflict of interest

The author has no conflict of interest.

Appendix

The parameters of the three motors taken as examples are presented below.

The DC motor: P_n = 1 kW, U_n = 220 V, n_n = 3000 rot./min., η = 0,75, p = 2, J = 0,006 kgm^2, M_n = 3,2 Nm, Ω_n = 314 rad/s, I_n = 6A, I_M = 10,8 A, R_a = 2,01 Ω, L_a = 0,034 H, T_a = 0,017 ms, k_e = 0,664 Vs, k_m = 0,533 Nm/A, k_f = 8.10^{-4} Nms, K_{CNA} = K_{CAN} = 1, I_{lim} = I_M, U_{aM} = 240 V.

The induction motor: R_s = 12,4 Ω; R_r = 12,4 Ω; L_m = 0,8 H; $L_{s\sigma}$ = 0,06 H; $L_{r\sigma}$ = 0,06 H; p = 4; k_f = 0,008 Nms; J = 0,01 kgm^2; M_{Mc} = 7 Nm; M_{Mt} = 24 Nm; n_b = 750 rot/min; $P_{N(Mc)}$ = 550 W; I_{sN} = 1,77 A; I_{sM} = 8 A; U_{sN} = 220 V.

The permanent magnet synchronous motor: P_N = 400 W; I_N = 3 A; I_M = 8 A; n_m = 4000 rot/min; n_b = 3000 rot/min; M_{Mc} = 1,3 Nm; M_{Mt} = 3,4 Nm; J = 0,001 kgm^2; k_f = 0,0001 Nms; R_s = 0,6 Ω; L_d = 4 mH; L_q = 5 mH; p = 4; Φ_{e0} = 0,072 Wb; V_{cc} = 200 V.

Author details

Constantin Volosencu
Department of Automation and Applied Informatics, "Politehnica" University Timisoara, Romania

*Address all correspondence to: constantin.volosencu@aut.upt.ro

References

[1] Sekar A. Fuzzy PI DC motor Speed Control. 2021. Available from: https://www.mathworks.com/matlabcentral/fileexchange/74291-fuzzy-pi-dc-motor-speed-control. MATLAB Central File Exchange. The Mathworks, Inc., Natick, MA, USA. [Accessed 2021-01-06]

[2] Malla S. Fuzzy controller based Speed Control of DC Motor. 2021. Available from: https://www.mathworks.com/matlabcentral/fileexchange/36508-fuzzy-controller-based-speed-control-of-dc-motor. MATLAB Central File Exchange. The Mathworks, Inc., Natick, MA, USA. [Accessed 2021-01-06]

[3] Merah A. Speed control of DC motor using Fuzzy Logic Controller. 2021. Available from: https://www.mathworks.com/matlabcentral/fileexchange/64993-speed-control-of-dc-motor-using-fuzzy-logic-controller. MATLAB Central File Exchange. The Mathworks, Inc., Natick, MA, USA. [Accessed 2021-01-06]

[4] Elbesealy M. A Fuzzy V/F Control for High Performance Induction Motors Drive. 2021. Available from: https://www.mathworks.com/matlabcentral/fileexchange/59454-a-fuzzy-v-f-control-for-high-performance-induction-motors-drive. MATLAB Central File Exchange. The Mathworks, Inc., Natick, MA, USA

[5] Sang N.T. Fuzzy controller induction motor. 2021. Available from: https://www.mathworks.com/matlabcentral/fileexchange/68377-fuzzy-controller-inductin-motor. MATLAB Central File Exchange. The Mathworks, Inc., Natick, MA, USA. [Accessed 2021-01-06]

[6] Zadeh LA. Fuzzy Logic Toolbox User's Guide. The Math Works, Inc., Natick, MA, USA.; 1995-2020.

[7] Sybille G, Dessaint LA. et al. Simscape Electrical Reference (Specialized Power Systems). Hydro-Québec and The MathWorks, Inc., Natick, MA, USA; 1998–2020

[8] Sybille G, Dessaint LA. et al. Simscape Electrical Reference. The MathWorks, Inc., Natick, MA, USA; 2013–2020

[9] Bose BK. Expert System, Fuzzy logic and neural network applications in power electronics and motion control. Proceedings of the IEEE. 1994;**8**: 1303-1323. DOI: 10.1109/5.301690

[10] Bolognani S, Zigliotto M. Fuzzy logic control of a switched reluctance motor drive. In: Conference Record of the 1993 IEEE Industry Applications Conference Twenty-Eighth IAS Annual Meeting; 2–8 October 1993; Toronto. Ontario, Canada: IEEE; 2005. DOI: 10.1109/IAS.1993.299145

[11] Cerruto E, Consoli A, Raciti A, Testa A. Fuzzy adaptive vector control of induction motor drives. IEEE Trans. on Power Electronics. 1997;**6**:1028-1040. DOI: 10.1109/63.641501

[12] Slemon GR. Electrical machines for variable-frequency drives. Proc. of the IEEE. 1994;**8**:1123-1139. DOI: 10.1109/5.301681

[13] Lorenz RD, Lipo TA, Novotny DW. Motion control with induction motors. Proc. of the IEEE. 1994;**8**:1215-1240. DOI: 10.1109/5.301685

[14] Jahnd TM. Motion control with permanent magnet AC mashines. Proc. of the IEEE. 1994;**8**:1241-1252. DOI: 10.1109/5.301686

[15] Holtz J. Pulse width modulation for electronic power conversion. Proc. of the IEEE. 1994;**8**:1194-1214. DOI: 10.1109/5.301684

[16] Mohan N, Robbins WP, Underland TM, Nilsen R, Mo O.

Simulation of power electronic and motion control systems - an overview. Proc. of the IEEE. 1994;**8**:1287-1302. DOI: 10.1109/5.301689

[17] Buhler H. Reglage par logique floue. Lausanne: Press Polytechniques et Universitaires Romands; 1994. p. 20

[18] Volosencu C. Fuzzy control of electrical drives, based on fuzzy logic. WSEAS Transactions on Systems and Control. 2008;**9**(3):809-822

[19] Volosencu C. Speed Control of a Permanent Synchronous Motor, Based on a Quasi-Fuzzy Controller, In Proceedings of the 10th European Conference on Power Electronics and Applications, European Power Electronics and Drives Association, 2003, Toulouse, France.

[20] Volosencu C. Some Consideration about Using Fuzzy Logic for Speed Control of Electrical Drives, In The 7-th European Congress on Intelligent Techniques and Soft Computing EUFIT'99, 1999. Aachen, Verlag Mainz, Germany

[21] Volosencu C. Demonstration by Simulation of the Robustness of the Fuzzy Control System of a DC Motor at Errors at the Parameter Identification, In Pemc'98 Conference, 1998, Prague, Czech Technical University, Czech Rep.

[22] Volosencu C. Introductory Chapter: Basic Properties of Fuzzy Relations, In Volosencu C, editor. Fuzzy Logic, IntechOpen, London, UK; 2020. p. 3-10. DOI: 10.5772/intechopen.88172

[23] Volosencu C. Properties of fuzzy systems. WSEAS Transactions on Systems. 2009;**2**(8):210-228

[24] Volosencu C. On Some Properties of Fuzzy Systems, In Proceedings of the 8th WSEAS International Conference on Signal Processing, Robotics and Automation (ISPRA09), Feb. 21–23, 2009; Cambridge, UK, 2005.

[25] Volosencu C. Tuning Fuzzy PID Controllers, In Rames C. Panda, editor. Theory, Tuning and Application to Frontier Areas, IntechOpen, London, UK.; 2012. 171-190. DOI: 10.5772/32750.

[26] Volosencu C. Pseudo-equivalence of fuzzy PID controllers. WSEAS Transactions on Systems and Control. 2009;**4**(4):163-176

[27] Volosencu C. Pseudo-Equivalence of Fuzzy PI Controller, In Proceedings of the 8th WSEAS International Conference on Signal Processing, Robotics and Automation (ISPRA09), Feb. 21–23; Cambridge, UK, 2009, WSEAS Press, Athens, Greece.

[28] Volosencu C. Stabilization of fuzzy control systems. WSEAS Transactions on Systems and Control. Athens, Greece: WSEAS Press; 2008;**10**(3): 879-896

[29] Leonhard W. Control of Electrical Drives. Berlin: Springer Verlag; 1985

[30] Boldea I, Nasar SA. Vector Control of A.C Drives. Boca Raton, Florida: C. R. C. Press; 1992